HANDMADE GIFT CARDS

the ultimate step-by-step guide to creating

HANDMADE
GIFT CARDS

75 imaginative designs to make, shown in
500 stunning photographs

practical projects for beautiful and original cards, gift wrap,
boxes, tags, wallets and invitations for every occasion

cheryl owen

southwater

This edition is published by Southwater, an imprint of Anness Publishing Ltd
Hermes House, 88–89 Blackfriars Road, London SE1 8HA
tel. 020 7401 2077; fax 020 7633 9499
www.southwaterbooks.com; www.annesspublishing.com

If you like the images in this book and would like to investigate
using them for publishing, promotions or advertising, please visit
our website www.practicalpictures.com for more information.

UK agent: The Manning Partnership Ltd
tel. 01225 478444; fax 01225 478440
sales@manning-partnership.co.uk

UK distributor: Book Trade Services
tel. 0116 2759086; fax 0116 2759090
uksales@booktradeservices.com
exportsales@booktradeservices.com

North American agent/distributor: National Book Network
tel. 301 459 3366; fax 301 429 5746
www.nbnbooks.com

Australian agent/distributor: Pan Macmillan Australia
tel. 1300 135 113; fax 1300 135 103
customer.service@macmillan.com.au

New Zealand agent/distributor: David Bateman Ltd
tel. (09) 415 7664; fax (09) 415 8892

Publisher: Joanna Lorenz
Editorial Director: Helen Sudell
Editor: Simona Hill
Designer: Ian Sandom
Photographers: Mark Wood and Paul Bricknell
Editorial Reader: Penelope Goodare
Production Controller: Mai-Ling Collyer

ACKNOWLEDGEMENTS
The author would like to thank Lucinda Ganderton for supplying the projects
on pages: 10, 13, 14, 21, 58, 59, 63, 79.

ETHICAL TRADING POLICY
At Anness Publishing we believe that business should be conducted in an
ethical and ecologically sustainable way, with respect for the environment and
a proper regard to the replacement of the natural resources we employ.
As a publisher, we use a lot of wood pulp to make high-quality paper for
printing, and that wood commonly comes from spruce trees. We are
therefore currently growing more than 750,000 trees in three Scottish forest
plantations: Berrymoss (130 hectares/320 acres), West Touxhill (125
hectares/305 acres) and Deveron Forest (75 hectares/185 acres). The forests
we manage contain more than 3.5 times the number of trees employed each
year in making paper for the books we manufacture.
Because of this ongoing ecological investment programme, you, as our
customer, can have the pleasure and reassurance of knowing that a tree is
being cultivated on your behalf to naturally replace the materials used to
make the book you are holding.
Our forestry programme is run in accordance with the UK Woodland
Assurance Scheme (UKWAS) and will be certified by the internationally
recognized Forest Stewardship Council (FSC). The FSC is a non-government
organization dedicated to promoting responsible management of the world's
forests. Certification ensures forests are managed in an environmentally
sustainable and socially responsible way. For further information about this
scheme, go to www.annesspublishing.com/trees

Previously published as part of a larger volume,
The Complete Practical Guide to Card-Making

PUBLISHER'S NOTE
Although the advice and information in this book are believed to be
accurate and true at the time of going to press, neither the authors nor
the publisher can accept any legal responsibility or liability for any errors
or omissions that may be made nor for any inaccuracies nor for any
harm or injury that comes about from following instructions or advice
in this book.

CONTENTS

INTRODUCTION

Greetings cards have a wide appeal across many cultures. We send them to mark special occasions such as birthdays or Christmas, to keep in touch, and to express goodwill. In sending these tokens to family and friends we are subscribing to a long tradition. Greetings have been sent in written form for hundreds of years. Originally greetings cards were the preserve of the rich, who commissioned them as elaborate gifts that were expensively produced. It was not until printing methods were modernized and printers developed the ability to reproduce colours accurately and at low cost that the market proliferated with cards for Christmas and Valentine's Day. In recent years cards have been produced for every occasion imaginable, as well as plain pictorial cards, providing an easy way to communicate with friends. A small and inexpensive gesture, a greetings card can be kept and treasured.

The craft of making greetings cards has grown in popularity, alongside the wider availability of specially designed craft materials, beautiful handmade papers, and the internet as a means of exchanging ideas. As handcrafted cards form a larger part of stationery ranges in gift shops, and are perceived to be of a high value, so the fashion for designing and making these items of value has blossomed in popularity.

Choosing to make and send a greetings card that is handcrafted to suit the style, interests and specific occasion to be celebrated by the recipient makes the gift more personal. It shows how much you care, and the receiver will appreciate the thought and effort that has gone into making the card. Greetings cards can be quickly made with minimal materials and financial outlay. Because they have a relatively small surface area to decorate, they are an ideal craft for people with little time to spare for creative pursuits, or who enjoy the pleasure and satisfaction of dabbling with lots of different craft skills.

This beautiful book presents 75 unique greetings card ideas for every landmark occasion, such as Christmas, Easter, wedding anniversaries, engagements and the arrival of a new baby, as well as cards to send just to keep in touch. There are projects for beginners as well as plenty of inspirational ideas for accomplished card makers. Many are simple enough to make quickly, and in multiples, so are suitable to make for invitations or for Christmas cards, for example, where a lot of one design can be made up following a template. Or you could lavish time and care to create a one-of-a-kind card for a special person.

A wide range of design styles are presented, to cater for every taste. There are retro ideas, such as a photo and cross-stich card that sets a favourite old photograph against a 1950s-style background for a charming nostalgic effect. Christmas cards with traditional motifs sit alongside a contemporary card for Diwali. There are cute and colourful cards for children's birthdays, as well as stylish and sophisticated cards for female relatives, and sport-themed cards for men.

Each card is illustrated with step-by-step photographs and there are comprehensive instructions to help you achieve the best result possible. Many of these designs can be made from everyday materials that are easy to locate and inexpensive to purchase. Templates are provided for every card where needed, so no freehand drawing is required. Once you have mastered new crafts for card making, be inventive and create your own innovative designs, that friends are sure to treasure.

Holiday celebrations

Fabulous festive occasions are annual highlights in the lives of people all over the world. For some celebrations, such as Valentine's Day, Mother's Day and Father's Day, you will have a particular loved one in mind and can handcraft a personal card just for them. You may want to mark other times of religious, spiritual or patriotic significance by sending large numbers of cards which may need to be more general in appearance and simpler to create. This chapter covers all these options, with many ideas for elaborate one-off creations as well as cards that can be made easily following a template.

New Year celebration

Capture the exuberant tone of New Year festivities with a mass of colourful balloons. As well as the balloon motifs stuck on the front of the card, balloon-shaped windows reveal shiny acetate in a design of depth and varied textures.

materials and equipment

- tracing paper and pencil
- scissors
- coloured paper in several shades and weights
- craft knife
- cutting mat
- white card blank
- coloured acetate
- glue stick
- star stickers

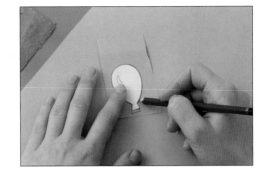

1 Trace the balloon templates at the back of the book and cut out the shapes. Draw around these with a sharpened pencil on to the various coloured papers so that you have seven different coloured balloons.

2 Cut out the shapes with scissors and then cut out the highlight shapes on each balloon using a craft knife and working on a cutting mat.

3 Using the templates, draw three balloon shapes on the front of the card blank and cut them out with the craft knife. Cut a rectangular piece of acetate, slightly smaller than the card, and attach it to the inside of the card front using a glue stick. Cut out the three highlight shapes from the offcuts of white card and glue them in position on the acetate balloons.

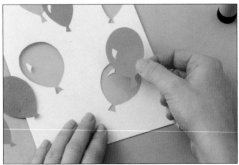

4 Arrange the coloured paper balloons across the card and, when you are pleased with the design, glue them in place. Finish off by adding a few twinkling star stickers.

Winged heart Valentine

The message is clear as a winged heart leaps out of this valentine card on a wire spring. The spring is easily made by wrapping fine wire around a pencil.

materials and equipment

- card (stock) in turquoise, red holographic hearts and gold
- craft knife
- metal ruler
- cutting mat
- bone folder
- tracing paper
- pencil
- 5mm/¼in wide double-sided tape
- 30cm/12in length of 0.4mm gold wire
- clear adhesive tape
- white paper

1 Cut a 41.5 x 14cm/16¼ x 5½in rectangle of turquoise card. On the wrong side, score and fold the card parallel with the short edges, 13.5cm/5¼in from the left edge and 14cm/5½in from the right edge. Open out. Draw a 6.5cm/2½in square on the central section, 2.5cm/1in from the top edge and right foldline. Cut out.

2 Use the template at the back of the book to cut a heart from red holographic hearts card and the wings from gold card. Stick the wings to the back of the heart using double-sided tape.

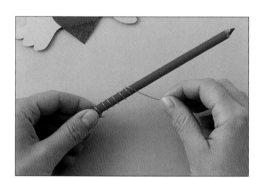

3 Wind the wire around a pencil. Slip the coil off the pencil and pull it slightly open. Stick one end of the wire to the back of the heart using clear adhesive tape.

4 Apply double-sided tape around the window on the wrong side. Stick the end of the wire coil behind the window with adhesive tape. Cut an 8.5cm/3½in square of white paper. Peel the backing strip off the double-sided tape and stick the square behind the window. Stick the facing behind the front using double-sided tape.

Valentine shoe

This fabulous shoe is decorated with a spray of colourful hearts. The shoe and card are made of beautiful printed papers that have the look of silk fabrics.

materials and equipment

- deep pink printed paper
- card (stock) in white, purple, deep pink, mauve, light green and pale pink
- purple printed paper
- spray adhesive
- craft knife
- metal ruler
- cutting mat
- bone folder
- white fibrous paper
- tracing paper
- pencil
- 0.4mm purple wire
- wire snippers
- glue dots

3 Cut a total of five small hearts from deep pink, mauve, light green and pale pink card and a large heart from deep pink card. (The hearts can be punched using heart-shaped paper punches.) Snip six 4cm/1½in lengths of purple wire. Stick each wire to the back of a small heart with a glue dot.

1 Apply deep pink printed paper to white card and purple printed paper to purple card using spray adhesive. Cut a 25 x 18cm/10 x 7in rectangle of white covered card using a craft knife and metal ruler. Score and fold it across the centre, parallel with the short edges, using a bone folder.

2 Cut a 13 x 7.5cm/5 x 3in rectangle of white fibrous paper. Stick the paper centrally to the front of the greetings card using spray adhesive. Use the template at the back of the book to cut out a shoe from the purple card and stick it centrally on the card front using spray adhesive.

4 Stick the other ends of the wires to the back of the large heart with glue dots, trimming some wires so that the small hearts are at different levels. Stick the large heart to the shoe using glue dots.

Victorian Valentine

Valentine cards were among the first commercially produced greetings cards in the mid-19th century, and many early examples were lavishly trimmed with coloured scraps and paper lace. This gorgeous confection echoes that style.

materials and equipment

- 2 paper doilies
- scissors
- gold tissue paper
- spray adhesive
- pink paper
- 30 x 25cm/12 x 10in rectangle of marbled lilac card (stock)
- selection of reproduction scraps, including cherubs, hearts and flowers
- 1 gold doily
- glue stick
- paper insert

1 Choose a round motif, about 12.5cm/5in in diameter, from the centre of one of the doilies. Cut it out roughly, then cut a piece of gold tissue slightly larger all round. Stick the tissue to the back of the doily using spray adhesive, then cut out the motif neatly around the edge.

2 Select a pair of motifs, each about 5 x 12.5cm/2 x 5in, from the other doily. Back these motifs with pink paper as before.

3 Fold the card in half widthways and sharpen the crease with a bone folder to give a crisp finish. Glue the round motif centrally to the front of the card, then stick one of the narrow motifs on each side. Glue a large scrap to the centre of the round motif then add the flowers, heart and cherubs.

4 Carefully snip a few flower shapes from the gold doily and add them to the arrangement. Glue in a paper insert inscribed with your own special message and decorated with another cherub.

Millefiori Easter cards

The clay egg motifs on these stylish Easter cards are decorated using the millefiori technique: different coloured clays are stretched and rolled together into multicoloured "canes", then sliced and embedded in the clay surface.

materials and equipment

- polymer clay in white, apricot, lilac and violet
- rolling pin
- chopping board
- egg-shaped cookie cutter
- craft knife
- 8mm/³⁄₈in-wide ribbons in toning shades
- scissors
- all-purpose household glue
- textured card blanks
- adhesive foam pads

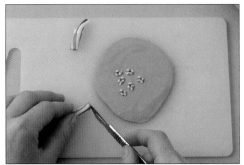

3 Gently roll the cane with the tips of your fingers until it is about 8mm/⅓in in diameter. Use a craft knife to cut off 2mm/¹⁄₁₂in slices. Arrange these over the surface of the clay oval, then roll it gently until the slices sink into the surface of the clay and no joins are visible.

4 Cut out an egg shape with the cookie cutter. Experiment with the remaining clay to make more millefiori patterns for other cards. Bake the egg motifs to harden them following the clay manufacturer's instructions and leave to cool. To complete each card, tie a 15cm/6in length of ribbon into a small bow, trim the ends and stick to the egg using all-purpose household glue. Attach the egg to the card front using adhesive foam pads.

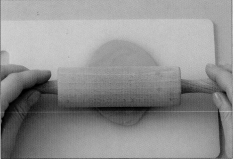

1 Cut off a quarter of one clay block and knead it until malleable. Roll it out on a chopping board to a depth of 4mm/⅛in, forming an oval shape slightly larger than the egg-shaped cookie cutter.

2 Cut and knead 2cm/¾in squares of the remaining three clay blocks. Roll each one out into a 4mm/⅛in sausage. Cut one length of one colour and three each of the other two. Arrange these lengths alternately around the single colour and press gently together to make a cane.

Easter egg card

Choose papers in subtle colours to make this unusual Easter card. The eggs are made from scraps of co-ordinating gift wrap and tied with fine string. Keep the colours neutral for a sophisticated look.

materials and equipment

- grey mulberry paper
- grey card (stock)
- spray adhesive
- craft knife
- metal ruler
- cutting mat
- bone folder
- crinkled copper paper
- pinking shears
- white writing paper with embedded fibres
- scraps of 3 different but co-ordinating gift wraps
- tracing paper
- pencil
- fine hemp string or embroidery thread (floss)
- scissors
- adhesive foam pads

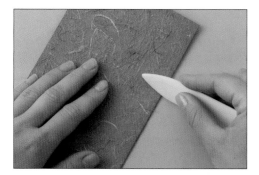

1 Apply grey mulberry paper to grey card using spray adhesive. Cut a 22 × 20.5cm/ 8½ × 8in rectangle of the covered card using a craft knife and metal ruler and working on a cutting mat. Score and fold the card across the centre, parallel with the short edges, using a bone folder.

2 Cut a 16.5 × 7cm/6½ × 2¾in rectangle of crinkled copper paper, using pinking shears to create a decorative edge. Cut a 15 × 5.5cm/6 × 2¼in rectangle of white writing paper with embedded fibres. Stick the copper paper centrally to the card front using spray adhesive, then attach the white paper in the centre of the copper panel in the same way.

3 Apply scraps of three gift wrap papers to offcuts of the grey card using spray adhesive. Use the egg template at the back of the book to draw matching egg shapes on each design and cut them out using a craft knife.

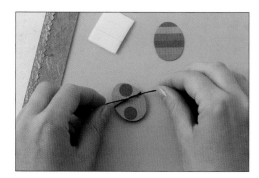

4 Tie a length of hemp string around each egg in a double knot and trim the ends. Alternatively, tie the eggs with embroidery thread (floss). Stick the eggs in a row to the card using adhesive foam pads.

Mother's Day gift

Here is a thoughtful gift for Mother's Day of a pretty book covered with a vintage scarf. Old silk scarves eventually fray at the edges, but their beautiful patterns and textures can be enjoyed again when recycled in such a way. The book has pages of handmade paper and a traditional pamphlet binding.

materials and equipment

- white card (stock)
- craft knife
- metal ruler
- cutting mat
- bone folder
- fabric scissors
- vintage scarf
- PVA (white) glue
- speckled cream paper
- spray adhesive
- 4 sheets of cream handmade A4 paper
- bradawl
- large-eyed needle
- 60cm/24in of 3mm/^{1}/$_8$in-wide satin ribbon

1 Cut a 31 × 22cm/12^1/$_4$ × 8^5/$_8$in rectangle of white card for the cover of the book, using a craft knife and metal ruler and working on a cutting mat. Score and fold the card across the centre, parallel with the short edges, using a bone folder. Open the card out flat.

2 Using fabric scissors, cut a 34 × 25cm/13^1/$_2$ × 10in rectangle from a vintage scarf, positioning the best part of the design on the right-hand side. Place the fabric wrong side up then centre the card cover on top. Stick the fabric corners on one short edge to the card cover with PVA glue, then stick the short edge to the cover.

3 Fold the cover in half. Adjust the fabric so that it lies smoothly. Lift the front cover and stick the remaining corners of the fabric inside the back cover using PVA glue. Stick the short edge inside the back cover.

4 Open the cover. Turn in the long fabric edges and stick them to the cover using PVA glue. Leave to dry.

5 Cut a 30.5 × 21.5cm/12 × 8^3/$_8$in rectangle of speckled cream paper. Centre the paper on the inside of the cover and stick in place using spray adhesive to conceal the raw edges of the fabric.

6 Stack the sheets of A4 handmade paper on a cutting mat. Place the cover centrally on top of the sheets.

7 Pierce a hole through all the layers at the centre of the central fold, using a bradawl. Pierce two more holes, 5cm/2in to each side of the centre.

8 Thread a large-eyed needle with ribbon. Starting on the outside of the cover, insert the needle through the centre hole then bring it out through the hole to one side of the centre. Take the ribbon through the hole on the other side of the centre and then out through the centre. Adjust the ribbon ends so that they are level and tie in a bow around the threaded ribbon. Fold the book in half.

Mother's Day appliqué

Here is an elegant flamingo of textured silk. Its eye and long legs are drawn with pink relief paint and its plumage is delicately highlighted with sequins.

materials and equipment

- white on cream printed paper
- cream card (stock)
- spray adhesive
- craft knife
- metal ruler
- cutting mat
- bone folder
- tracing paper and pencil
- 11cm/4$\frac{1}{2}$in square of iron-on interfacing
- 11cm/4$\frac{1}{2}$in square of pink silk dupion
- iron
- fabric scissors
- pink relief paint
- pink sequins
- glue dots
- tweezers

3 Stick the flamingo to the card front using spray adhesive. Draw the bird's legs on the card lightly with a pencil. Use pink relief paint to draw an eye and redraw the legs. Set aside to dry.

1 Stick white on cream printed paper to cream card. Cut a 20cm/8in square of the covered card using a craft knife and metal ruler and working on a cutting mat. Score and fold it across the centre.

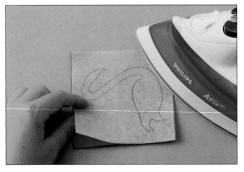

2 Trace the flamingo template at the back of the book on to the rough side of iron-on interfacing. Place the interfacing, adhesive side down, on a piece of pink silk dupion and iron it on. Cut out the flamingo.

4 Stick a few pink sequins to the flamingo with glue dots, positioning them on the wing tips and tail with tweezers.

Golfing Father's Day card

If your dad is a keen golfer this is the card to make for him. Coloured bands are painted across the card and wooden golf tees are slotted through velvet ribbon – they can be slipped out later to use on the golf course.

materials and equipment

- white card (stock)
- craft knife
- metal ruler
- cutting mat
- bone folder
- scrap paper
- 2.5cm/1in-wide masking tape
- acrylic paint in gold, green and smoky blue
- flat paintbrush
- 20cm/8in of 1cm/³⁄₈in-wide ridged smoky blue velvet ribbon
- 4 wooden golf tees
- double-sided tape
- embroidery scissors

1 Cut a 23 × 19cm/9 × 7½in rectangle of white card. Score and fold the card across the centre, parallel with the short edges, using a bone folder. Resting on scrap paper, apply two lengths of masking tape across the card front, to mask off a band 7cm/2¾in wide, 4.5cm/1¾in below the upper edge. (Test the tape on a scrap of the card first to check it doesn't damage the surface.)

2 Stick two lengths of masking tape lightly to a cutting mat and cut two 5mm/¼in-wide and two 1cm/³⁄₈in-wide strips. Apply each to the card front above and below the masked band. Paint the card between the tapes, leaving the wide band clear. Peel off the masking tape before the paint dries.

3 When the paint is dry, open the card out and cut four pairs of 1.2cm/½in vertical slits, 1cm/³⁄₈in apart, across the wide band. They should be 1.5cm/⅝in in from the sides, with 1.5cm/⅝in gaps between the pairs of slits.

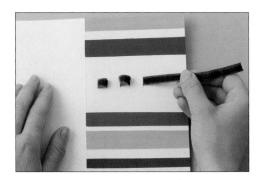

4 Thread velvet ribbon in and out of the slits. Slip a golf tee behind each ribbon loop. Stick the ends of the ribbon to the underside of the card front with double-sided tape. Snip off the excess ribbon.

Greetings for Diwali

Wish family and friends peace and happiness for the festival of lights with this celebratory Diwali card depicting a lit diya.

materials and equipment

- red and yellow card (stock)
- craft knife
- metal ruler
- cutting mat
- bone folder
- tracing paper and pencil
- lilac pearlized card (stock)
- gold paper
- spray adhesive
- double-sided tape
- 7 x 6mm/¹/₄in red round glitter stickers
- 4 x 8mm/⁵/₁₆in purple teardrop glitter stickers

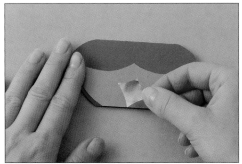

3 Use the template to cut the diya front from lilac pearlized card, the outer flame from yellow card and the diamond and inner flame from gold paper. Stick the diya front to the card front, the diamond to the diya front and the inner flame to the outer flame using spray adhesive.

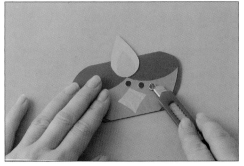

4 Stick the outer flame to the card front using double-sided tape. Stick red round stickers along the upper edge of the diya front and decorate the diamond with purple teardrop stickers, positioning them with the blade of the craft knife.

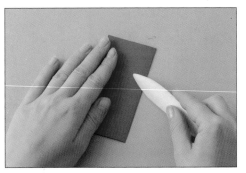

I Cut an 11cm/4⅜in square of red card using a craft knife and metal ruler and working on a cutting mat. Score and fold the card across the centre using a bone folder.

2 Use the template at the back of the book to draw the diya on the card front, with the fold along the top. Resting on a cutting mat, cut out the diya.

Hanukkah greeting

The seven-branched candelabrum, called a menorah, is associated with Hanukkah. Create this distinctive symbol using relief paint to send good wishes.

materials and equipment

- maroon and white card (stock)
- craft knife and metal ruler
- cutting mat
- tracing paper
- pencil
- bradawl
- gold relief paint
- 7 x 3mm/^{1}/$_8$in square diamanté stickers
- embroidery needle
- silver thread
- 7 x 10mm/^{3}/$_8$in silver bugle beads
- gold glitter paint
- bone folder
- spray adhesive

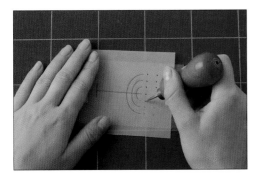

1 Cut a 9.5 x 7cm/3¾ x 2¾in rectangle of maroon card using a craft knife and metal ruler and working on a cutting mat. Use the template at the back of the book to draw the menorah in the centre of the rectangle. Resting on a cutting mat, pierce a hole at each dot using a bradawl.

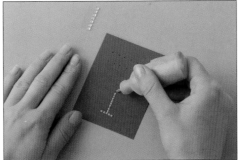

2 Dot gold relief paint along the drawn lines and set aside to dry. Stick a 3mm/⅛in square diamanté sticker at the top of each branch of the menorah.

3 Thread an embroidery needle with a double length of silver thread. Knot the ends together and sew a silver bugle bead between each pair of holes to resemble candles. Apply gold glitter paint above each bead to look like flames. Set aside to dry.

4 Cut a 22 x 14.5cm/8⅝ x 5¾in rectangle of white card. Score and fold the card across the centre, parallel with the short edges, using a bone folder. Glue the maroon rectangle to the centre of the card front using spray adhesive.

Halloween card

Here is a fun idea for a Halloween card. A scary foam spider bounces on a length of elastic, staring eerily from a pair of joggle eyes.

materials and equipment

- bright green mottled card (stock)
- craft knife
- metal ruler
- cutting mat
- bone folder
- tracing paper
- pencil
- black Neoprene foam
- metallic pen
- two 6mm/¼in joggle eyes
- glue dots
- bradawl
- 15cm/6in fine black round elastic

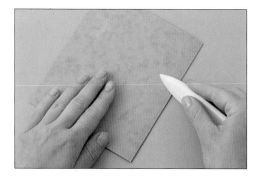

1 Cut a 26 × 16cm/10¼ × 6¼in rectangle of bright green mottled card using a craft knife and metal ruler and working on a cutting mat. Score and fold the card across the centre, parallel with the short edges, using a bone folder.

2 Use the template at the back of the book to draw a spider on black Neoprene foam using a metallic pen. Cut out the spider using a craft knife on a cutting mat. Stick a pair of joggle eyes on the spider using glue dots.

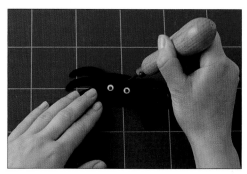

3 Pierce a hole in the centre of the spider with a bradawl, resting on a cutting mat. Thread elastic through the hole. Knot the end of the elastic under the spider.

4 Open the card out flat. Resting it on a cutting mat, pierce a hole in the card front using a bradawl, 1.5cm/⅝in below the centre of the upper edge. Insert the end of the elastic in the hole and knot the elastic inside the card.

Tip
When you are cutting shapes from dark-coloured Neoprene foam, a metallic pen is best for drawing lines as it shows up clearly.

Thanksgiving garland card

This pretty circlet in the rich colours of November in America is made using coloured paper leaves, but if you have planned ahead and have a store of colourful pressed real autumn leaves you could use them instead in a similar design.

materials and equipment

- tracing paper or three small leaves
- pencil
- scissors
- paper in shades of brown, green and orange
- fine knitting needle, crochet hook or large tapestry needle
- 18 x 30cm/7 x 12in duck-egg blue card (stock)
- bone folder
- glue stick

1 Trace the three leaf templates at the back of the book and cut them out. Alternatively find a few small leaves and use them as templates. Use the templates to draw 11 outlines on to the different coloured papers.

2 Cut out the leaf shapes with sharp scissors, carefully following the outlines.

3 Use the point of a knitting needle, crochet hook or tapestry needle to score a pattern of veins across the front of each leaf. Gently fold along the curves to give a three-dimensional look to the leaves.

4 Fold the card in half lengthwise and rub over the crease with a bone folder. Arrange the leaves in a circle on the card front, overlapping them to create a garland. Use a glue stick to fix them in place.

Thanksgiving notebook

Get all your loved ones to write their thoughts in this rustic notebook when you gather to celebrate Thanksgiving. The cover and pages are made from handmade papers and hinged with a twig and raffia. A pressed autumn leaf is a lovely simple motif to apply to the front cover.

materials and equipment

- autumn leaf
- blotting paper
- flower press or heavy book
- bone folder
- ruler
- 2 x A5 sheets of dark green handmade paper
- cutting mat
- 8 x A5 sheets of light green handmade paper
- twig, approximately 12.5cm/5in long
- bradawl
- large-eyed needle
- undyed raffia
- scissors
- A5 sheet of sage green handmade paper
- PVA (white) glue

1 Press a leaf between sheets of blotting paper in a flower press or a heavy book for about 10 days. Using a bone folder and ruler, score and fold an A5 sheet of dark green handmade paper 3cm/1¼in from one short edge. Open the sheet out flat again. This will be the front cover; the remaining sheet will be the back cover.

2 Resting on a cutting mat, stack eight A5 sheets of light green handmade paper on top of the back cover. Place the front cover on top with the scored fold to the left to form the hinge.

3 Lay the twig on the hinge. Using a bradawl, pierce two pairs of holes on each side of the twig at the top and bottom to make cross stitches.

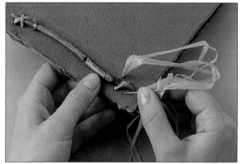

4 Thread a large-eyed needle with undyed raffia. Leaving a trailing end of raffia at the back, sew through the holes using cross stitches to anchor the twig to the hinge at the top and bottom.

5 Tie the ends of the raffia together securely on the back cover. Cut off the excess raffia with a pair of scissors.

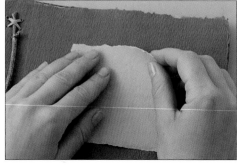

6 Tear a 9cm/3½in square of sage green handmade paper by tearing the paper against a ruler. Stick the square to the front cover of the book using PVA glue. Finally, stick the leaf on the square with PVA glue.

Dove of peace Christmas card

This three-dimensional paper dove would look lovely perched among the branches of a festive tree, but it will also stand on a flat surface, balanced by its tail, so it can be displayed on a shelf with other Christmas cards. Write your festive message on the back of the body and fold the bird flat for mailing.

1 Stick speckled cream paper to cream card using spray adhesive. Use the template at the back of the book to cut the bird's body from the covered card.

2 Cut out two 8cm/3¼in squares of speckled cream paper and fold them in half. Use the template at the back of the book to draw a wing and tail on each piece, matching the folds. Draw the feathers. Cut out the pieces using a craft knife, resting on a cutting mat.

materials and equipment

- speckled cream paper
- cream card (stock)
- spray adhesive
- tracing paper
- pencil
- craft knife
- cutting mat
- all-purpose household glue
- crystal jewellery stone sticker

3 Cut the notches representing the feathers with a craft knife, resting on a cutting mat. Open the wings and tail out flat. Cut the wing section in half to make a pair. Lift the feather tips outwards.

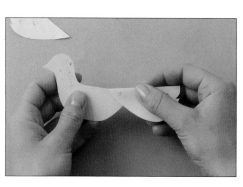

4 Dot all-purpose household glue on the underside of the tail near the point. Stick the tail over the end of the bird.

5 Dot all-purpose household glue on the undersides of the wings at the points. Stick the wings to each side of the bird.

6 Stick a jewellery stone sticker to the bird's head to represent an eye. To apply it, lift the sticker from its backing sheet with the tip of a craft knife blade. Place in position, remove the craft knife then press the sticker down firmly.

Metal plaque Christmas card

Here is a design that is ideal if you want to produce a set of Christmas cards, as it is very quick to make. Packs of metal plaques with Christmas motifs and an adhesive backing are available from craft suppliers.

1 Cut a 21 × 12cm/8¼ × 4¾in rectangle of pale yellow card using a craft knife and metal ruler and working on a cutting mat. Score and fold the card across the centre, parallel with the short edges, using a bone folder.

2 Peel the backing strip off the plaque and stick it to the centre of the card front.

materials and equipment

- pale yellow card (stock)
- craft knife
- metal ruler
- cutting mat
- bone folder
- 3.5 × 2.5cm/1³⁄₈ × 1in metal Christmas plaque
- pencil
- silver relief paint

3 In pencil, draw a simple swirling border on the card around each side of the plaque. Draw along the pencil lines with silver relief paint and leave to dry.

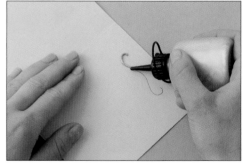

4 As a finishing touch, draw another swirl inside the card with a pencil. Draw along the pencil line with silver relief paint then leave the card open to dry.

Tip

Relief paints are sold in small bottles with thin nozzles, allowing them to be applied straight from the container. The technique takes a little practice, so try some swirls on scraps of card first so that your lines are confident and smooth.

Silver charms Christmas card

This vibrant Christmas card is easy to assemble using a few ready-made trimmings bought from craft stores. The metal charms are highlighted with dainty jewellery stones and attached to the card with metallic brads.

materials and equipment

- bright pink card (stock)
- craft knife
- metal ruler
- cutting mat
- bone folder
- star, snowflake and spiral metal charms
- tiny pink star and circular jewellery stone stickers
- bradawl
- 3 x 4mm/⁵/₃₂in pink brads
- light pink paper
- paper glue

1 Cut an 18 x 15cm/7 x 6in rectangle of bright pink card. Score and fold the card across the centre, parallel with the short edges, using a bone folder.

2 Decorate the charms with a few jewellery stone stickers. To apply, lift the stickers one at a time from their backing sheet with the tip of a craft knife blade. Place them on the charms, remove the craft knife then press the stickers in position.

3 Open the card out flat on a cutting mat. Arrange the charms on the card front. Use a bradawl to pierce holes in the card through the holes in the charms.

4 Insert a brad through each charm into the card and open the prongs inside. Cut a 17 x 14cm/6⅝ x 5½in rectangle of light pink paper for the insert. Fold it in half parallel with the short edges. Apply paper glue to the fold and insert, matching the folds.

Pressed ivy leaf gift tags

Press a number of ivy leaves to make a set of these natural gift tags, which look particularly good tied on parcels wrapped in brown paper.

materials and equipment

- ivy leaves
- blotting paper
- flower press or heavy book
- scrap paper
- copper metallic wax
- kitchen paper
- dark green and light green handmade paper
- ruler
- spray adhesive
- PVA (white) glue
- single hole punch
- yellow ochre hemp string
- scissors

1 Select unblemished ivy leaves and press them between sheets of blotting paper in a flower press or within the pages of a heavy book. Set aside for about 10 days.

2 Resting each leaf on a piece of scrap paper, rub copper metallic wax sparingly over the surface using kitchen paper.

3 Tear a 7.5cm/3in square of dark green handmade paper and a 6.5cm/2½in square of light green handmade paper, tearing the paper against a ruler. Stick the light green square centrally on the dark green square using spray adhesive.

4 Punch a hole in the top left corner of the gift tag using a 3mm/⅛in hole punch. Tie a length of hemp string through the hole for hanging. Stick the leaf to the front of the tag using PVA glue. Cut four 30cm/12in lengths of hemp string. Hold the lengths together and tie in a bow around the hanging string at the corner of the tag.

Organza tree gift tag

This charming festive gift tag is simple to make using offcuts of organza and a single bird-shaped sequin, so you can make enough for all your presents.

materials and equipment

- cream card (stock)
- craft knife
- metal ruler
- cutting mat
- bone folder
- fabric scissors
- scraps of gold and pink organza
- spray adhesive
- tracing paper
- pen
- bird-shaped sequin
- glue dot
- single hole punch
- 30cm/12in of 12mm/¹/₂in-wide gold organza ribbon

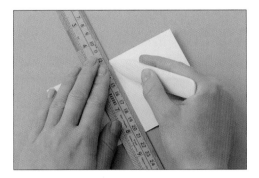

1 Cut a 14 × 8.5cm/5½ × 3⅜in rectangle of cream card using a craft knife and metal ruler and working on a cutting mat. Score the card across the centre, parallel with the short edges, using a bone folder, then fold in half and sharpen the crease with the side of the folder.

2 Using fabric scissors, cut a 6.5 × 5cm/ 2⅝ × 2in rectangle of gold organza. Stick the rectangle centrally to the front of the gift tag using spray adhesive.

3 Using the template at the back of the book, cut a tree from pink organza. Stick to the front of the tag using spray adhesive.

4 Stick a bird-shaped sequin to the tree using a glue dot. Punch a hole in the top left corner of the gift tag and tie a length of organza ribbon through the hole.

Birthdays

All age ranges are catered for in this chapter of birthday cards, from a first to a magnificent one hundredth. It is always fun to make cards for children, so there are plenty of ideas here using fantastic innovative materials. Many include numbers for specific ages, but you can of course substitute any number you want.

Adults love to receive birthday cards too, and there are plenty of themed cards for special interests such as surfing, rugby or music, plus lots of gift-wrapping suggestions to make your presents extra special.

Birthday cake gift tag

This professional-looking birthday cake is stamped with a rubber stamp, then coloured with felt-tipped pens and highlighted with glitter paint.

materials and equipment

- pale yellow and mid-blue card (stock)
- decorative-edged scissors
- birthday cake rubber stamp about 4cm/1½in square
- black ink pad
- felt-tipped pens in pale blue, pale pink and deep pink
- glitter paint in blue and gold
- craft knife
- metal ruler
- cutting mat
- bone folder
- spray adhesive
- single hole punch
- 25cm/10in of 10mm/³/₈in-wide yellow ribbon

1 Cut a 6.5 x 6cm/2⅝ x 2¼in rectangle of pale yellow card using a pair of decorative-edged scissors. Stamp a birthday cake rubber stamp on to a black ink pad. Stamp the image on to the centre of the yellow card. Leave the ink to dry.

2 Colour the stamped image using felt-tipped pens, colouring the plate pale blue, the cake pale pink and the cake band and candles deep pink.

3 Dot the cake with blue glitter paint and the candle flames with gold glitter paint. Set aside to dry.

4 Cut a 15 x 8cm/6 x 3¼in rectangle of mid-blue card using a craft knife and metal ruler and working on a cutting mat. Score and fold the card across the centre, parallel with the short edges, using a bone folder. Stick the pale yellow card to the front of the tag using spray adhesive. Punch a hole in the top left corner and fasten a length of ribbon through the hole.

Second birthday party invitation

These realistic gingerbread character cards make great party invitations. Use a cookie cutter as a template and imitate the traditional white icing with relief paint. A jaunty gingham ribbon around the neck completes the effect. The invitations are quick to make so you can easily produce a whole batch to send to the birthday person's friends.

materials and equipment

- gingerbread man cookie cutter
- pen
- terracotta-brown card (stock)
- white relief paint
- cutting mat
- craft knife
- 20cm/8in of 5mm/¼in-wide gingham ribbon
- fabric scissors

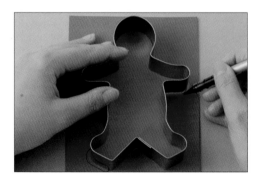

1 Draw around a gingerbread man cookie cutter on terracotta-brown card.

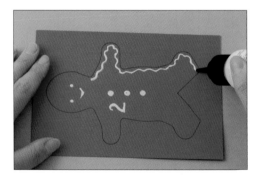

2 Decorate the character with white relief paint. Working out from the centre so that you do not smudge the paint, apply a row of buttons, a number 2, two eyes and a mouth, then outline the character with a wiggly line. Set aside to dry.

3 Resting on a cutting mat, cut out the character using a craft knife.

4 Tie a length of narrow gingham ribbon around the neck and trim the ends.

Concertina caterpillar

Masking fluid is applied to areas that you do not want to be painted. After paint has been applied, the dried fluid is rubbed away to reveal the unpainted paper beneath. It is ideal for the patterns on this friendly caterpillar.

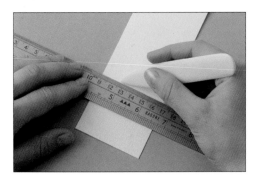

1 Cut a 33 × 6cm/13⅛ × 2⅜in strip of watercolour paper. Score it at 5.5cm/2³⁄₁₆in intervals, parallel with the short edges, using a bone folder. Fold the strip in accordion folds. Open the paper out flat again.

materials and equipment

- watercolour paper
- cutting mat
- craft knife
- metal ruler
- bone folder
- masking fluid
- medium artist's paintbrush
- watercolour paints in light green, lemon yellow and cerulean blue
- scrap paper
- tracing paper and pencil
- red relief paint

2 With masking fluid, paint a 2 on the second section from the right. Paint wavy lines and dots on the rest of the strip, leaving the right-hand section free.

4 Rub away the masking fluid with your finger to reveal the number and the patterns on the paper. Refold the strip.

5 Use the template at the back of the book to draw the shape of the caterpillar on the top section, matching the foldlines.

3 Leave the masking fluid to dry then rest the strip on scrap paper and paint it with vertical stripes of light green, lemon yellow and cerulean blue watercolour paints, blending the colours together on the paper. Leave to dry.

6 Cut through the layers using a craft knife. Draw the face on the front and outline the number using red relief paint. Leave to dry.

Stencilled birthday car

This card is bound to appeal to a young child who is keen on cars. The smart red vehicle has wheels that really turn and there are also three silver star stickers inside the card to mark a third birthday.

materials and equipment

- pale blue, dark blue and silver card (stock)
- craft knife
- metal ruler
- cutting mat
- bone folder
- tracing paper
- pen
- stencil board
- masking tape
- red and yellow acrylic paint
- stencil brush
- kitchen paper
- double-sided tape
- bradawl
- 2 x 12mm/$^1/_2$in diameter green star brads
- 3 x 2.5cm/1in diameter silver star stickers

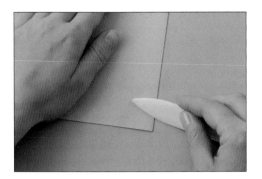

1 Cut a 22 × 16cm/8¾ × 6½in rectangle of pale blue card using a craft knife and metal ruler and working on a cutting mat. Score and fold the card across the centre, parallel with the short edges, using a bone folder.

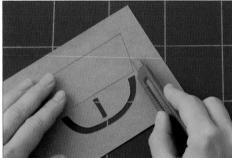

2 Trace the template at the back of the book and transfer the outlines of the car and the number 3 to stencil board. Cut out the two stencils using a craft knife, resting on a cutting mat.

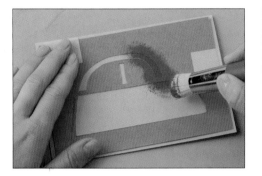

3 Tape the car stencil to the card front with masking tape. To stencil the car, pick up a small amount of red paint with a stencil brush and dab off the excess on kitchen paper. Dab the paint through the stencil, holding the brush upright and moving it in a circular motion. Leave to dry, then remove the stencil.

4 To judge the position of the number 3 stencil, hold the traced template over the card. Slip the number 3 stencil underneath, matching its position. Tape the stencil in place. Stencil the number with yellow paint. Remove the stencil.

5 Refer to the template to cut two wheels from dark blue card and two hub caps from silver card. Stick the hub caps to the wheels with small pieces of double-sided tape.

6 Open the card out flat on a cutting mat. Arrange the wheels on the car. Pierce a hole through the centre of each hub cap into the card using a bradawl.

7 Insert star-shaped brads in the holes in the hub caps and through the card. Splay open the prongs inside the card.

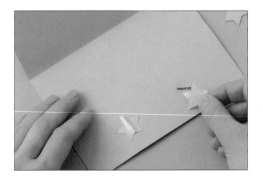

8 To protect small hands from the prongs, stick a star sticker on top of each brad inside the card. Stick on a third star at random inside the card.

Cute birthday dog

Delight a child with this comical dog. He has a real removable balloon for a tongue, so don't give this card to a child younger than four.

materials and equipment

- white and black card (stock)
- craft knife
- metal ruler
- cutting mat
- bone folder
- tracing paper
- pencil
- grey writing paper
- spray adhesive
- black medium felt-tipped pen
- double-sided tape
- red balloon

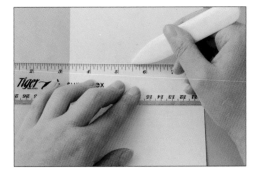

1 Cut a 23 × 12cm/9 × 4¾in rectangle of white card using a craft knife and metal ruler and working on a cutting mat. Score and fold the card across the centre, parallel with the short edges, using a bone folder.

2 Use the template at the back of the book to draw the dog's face on the folded card, leaving the fold to form the left-hand edge. Cut round the outline using a craft knife.

3 Roughly tear a patch of grey paper between your fingers. Stick the patch to the face using spray adhesive.

4 Lightly draw the facial features on the face with a pencil. Redraw the features with a black medium felt-tipped pen.

5 Resting on a cutting mat, use a craft knife to cut along the mouth just below the drawn line.

6 Use the template to cut a pair of ears from black card. Score and fold along the broken lines with a bone folder.

7 Glue the ears to the back of the card using double-sided tape and fold so that they drape over the front.

8 Carefully slip the balloon through the slit under the mouth and pull it inside the card until it resembles a dog's tongue.

Cowboy card

Nimble-fingered older children will enjoy making this dapper cowboy for their younger siblings. The various elements of the figure, his clothes and the age number are torn from a selection of plain and patterned papers.

materials and equipment

- fawn textured card (stock)
- craft knife
- metal ruler
- cutting mat
- bone folder
- tracing paper
- pencil
- masking tape
- salmon pink, black, blue, light brown and bright green plain paper
- green checked paper
- pale blue spotted paper
- black felt-tipped pen
- glue dots
- 12mm/½in silver holographic star sequin
- 2 silver star confetti
- paper glue

1 Cut a 26 × 19cm/10¼ × 7½in rectangle of fawn textured card using a craft knife and metal ruler and working on a cutting mat. Score and fold the card across the centre, parallel with the short edges, using a bone folder. Set aside.

2 Trace the cowboy templates at the back of the book and transfer the different pieces, right side down, to the wrong side of coloured and patterned papers as follows: head and hands on salmon pink paper; shirt on green checked paper, hat, moustache and boots on black paper, trousers and number 5 on blue paper, waistcoat on light brown paper and neckerchief on pale blue spotted paper. Leave a margin of at least 5mm/¼in around each image to make them easier to tear.

3 Tear all the pieces along the drawn lines by holding the image between the thumb and fingers of one hand close to the drawn line and pulling the paper around the image toward you with the other hand. Gradually move your fingers along the line.

4 Arrange all the pieces on the card front, being careful to place the shoes and hands below the trousers and shirt. Add the head after the shirt. Stick the pieces in place using spray adhesive.

5 Dot the eyes on the head with a black felt-tipped pen. Using glue dots, stick a silver holographic star sequin to the waistcoat and a confetti star to each boot as spurs.

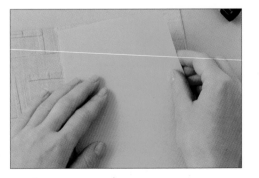

◄6 Cut a 25 × 18cm/9¾ × 7in rectangle of bright green paper for the insert. Fold the paper across the centre, parallel with the short edges. Run a line of paper glue along the fold and stick inside the card.

Birthday medallion card

Award a medal to your champion with this glittery card. The polymer clay medal is hung on ribbon and can easily be removed for the recipient to wear on their birthday, proclaiming their new age.

materials and equipment

- polymer clay in red, gold glitter and green glitter
- baking parchment
- drinking straw
- small kitchen knife
- 75cm/30in of 1cm/³/₈in-wide red ribbon
- fabric scissors
- blue glitter card (stock)
- craft knife
- metal ruler
- cutting mat
- bone folder

1 Roll a 2.5cm/1in ball of red polymer clay. On a sheet of baking parchment, flatten the clay to a 4.5cm/1¾in diameter circle for the medal. Stamp a hole at the top of the circle using a drinking straw.

2 Roll 1.5cm/⅝in balls of gold glitter and green glitter polymer clay. Roll the balls into two logs 11.5cm/4½in long on baking parchment. Twist the logs together.

3 Roll the twisted log until it is about 5mm/¼in thick, then bend the log into a number 6.

4 Carefully place the numeral on the red disc of clay and press it gently in place. Cut off the excess clay. Bake the medal to harden it following the clay manufacturer's instructions. Leave to cool.

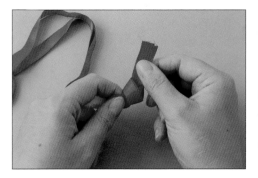

5 Thread a length of red ribbon through the hole in the medal. Knot the ribbon ends together and trim diagonally.

6 Cut a 24 x 16cm/9½ x 6¼in rectangle of blue glitter card using a craft knife and metal ruler and working on a cutting mat. Score and fold the card across the centre, parallel with the short edges, using a bone folder.

7 Open the card out flat. Cut two 2cm/¾in slits in the upper edge of the card front, 5cm/2in apart.

8 Slot the ribbon through the slits so the medal hangs on the front of the card.

Tip
If the blue glitter card sheds glitter, spray it with fixative spray before cutting it.

Russian dolls

This pretty design is based on traditional Russian matryoshka or nesting dolls, and if you wish you can make a trio of cards: reduce the size of the card to make a smaller one for another friend to send, and make an even smaller card for a gift tag, so the birthday person receives a complete set of dolls.

materials and equipment

- cream and red card (stock)
- craft knife
- metal ruler
- cutting mat
- bone folder
- tracing paper
- pencil
- spray adhesive
- double-sided tape
- salmon pink, black, turquoise and pink paper
- black, pale pink and red felt-tipped pens
- 12mm/$\frac{1}{2}$in tulip paper punch

1 Cut a 20 × 16.5cm/8 × 6½in rectangle of cream card using a craft knife and metal ruler and working on a cutting mat. Score and fold the card across the centre, parallel with the long edges, using a bone folder.

2 Trace the template at the back of the book and use it to draw the Russian doll on the folded card, aligning the lower left side of the doll with the fold. Use the template to cut a scarf, knot and feet from red card. Stick the scarf, then the knot, to the doll using spray adhesive. Stick the feet to the doll using double-sided tape.

3 Resting the card on a cutting mat, use a craft knife to cut out the doll, cutting through both layers and cutting around the right-hand curved edge of the feet.

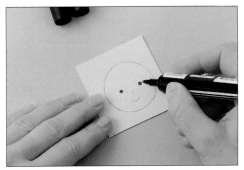

4 Use the template to draw the face lightly on salmon-pink card with a pencil. Dot the eyes with a black felt-tipped pen and the nose with a pale pink felt-tipped pen. Draw the mouth with a red felt-tipped pen.

5 Cut the hair from black paper and stick it to the face using spray adhesive. Cut out the face and stick it in the centre of the scarf using spray adhesive.

6 Cut the number 7 from turquoise paper. Use spray adhesive to stick the number to the card front.

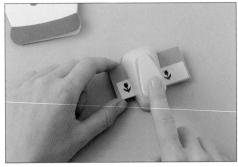

7 Apply double-sided tape to a strip of pink paper to make stickers to decorate the doll. Punch six tulips with a 12mm/½in tulip-shaped paper punch.

8 Peel off the backing papers and stick a row of three tulips down each side of the doll's dress, positioning them with a craft knife blade.

Butterfly birthday card

Transfer paint is a special non-toxic water-based paint that dries to form a soft, pliable film which can then be heat-transferred to many surfaces such as card and clothes. It can also be peeled off and reused. This transfer-painted butterfly has been applied to a turquoise card laced with ribbon and heart-shaped beads.

1 Trace the butterfly and the number 8 template at the back of the book on to tracing paper with a black pen. Tape a sheet of plastic on top of the tracing. Draw the butterfly using purple transfer paint.

2 Fill in the outline of the butterfly with transfer paints. Draw the numeral with red transfer paint. Set aside for 24 hours for the paint to dry.

3 Cut a 16.5 × 14.5cm/6½ × 5¾in rectangle of turquoise card, using a craft knife and metal ruler and working on a cutting mat, for the card front. Score and fold 2cm/¾in from one short side, using a bone folder, to form the hinge. Cut another rectangle of turquoise card measuring 19.5 × 14.5cm/ 7¾ × 5¾in for the back.

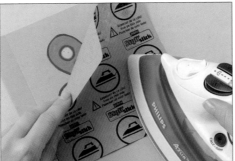

4 Place the card front on an ironing board, right side up with the hinge to the left. Carefully peel the designs off the plastic sheet. Position the butterfly on the card front. Place transfer paper on top. With the iron set to a silk setting, iron over the transfer paper. Do not use steam. Leave the motif to cool.

5 Place the card back on the ironing board, and place the front on top, matching the left-hand edges. Position the numeral on the right-hand edge of the back. Remove the front. Place transfer paper on the numeral and iron it on. Leave to cool.

6 Tape the front to the back, matching the left-hand edges. With a pen, draw a row of ten holes 1.5cm/⅝in apart along the centre of the hinge. Resting on a cutting mat, punch a hole at each dot using a 3mm/⅛in punch and a tack hammer.

materials and equipment

- tracing paper
- black pen
- sheet of plastic
- masking tape
- transfer paints in purple, silver, red, pink and yellow
- turquoise card (stock)
- craft knife
- metal ruler
- cutting mat
- bone folder
- transfer paper
- iron
- 3mm/⅛in hole punch
- tack hammer
- 30cm/12in of 3mm/⅛in-wide red ribbon
- 4 heart-shaped pony beads in assorted colours
- fabric scissors

7 Make a double knot at the end of the ribbon and thread it in and out of the punched holes, threading on a pony bead on each stitch at the front of the card. Make a double knot over the last hole and cut off the excess ribbon. Remove the tape. Enclose a piece of transfer paper with the card, with instructions telling an adult how to peel the motifs off the card and iron them on to the child's clothing by ironing over the transfer paper on a silk setting.

Funky foam monster birthday card

This friendly monster will bring a smile to the recipient's face. He is made from foam and proudly holds a number showing the child's age. The face and glitter spots are applied with relief paint.

materials and equipment

- yellow card (stock)
- craft knife
- metal ruler
- cutting mat
- bone folder
- tracing paper
- pencil
- pen
- Neoprene foam in red and aquamarine
- relief paint in lime green and silver
- all-purpose household glue
- glue dots

1 Cut a 26 x 17cm/10¼ x 6¾in rectangle of yellow card using a craft knife and metal ruler and working on a cutting mat. Score and fold the card across the centre, parallel with the short edges, using a bone folder.

2 Using the template at the back of the book cut out a monster from red Neoprene foam and a number 9 from aquamarine foam, using a craft knife and resting on a cutting mat.

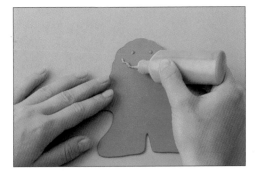

3 Draw a face on the monster using a pen. Redraw the lines with lime green relief paint. Set aside to dry.

4 Stick the numeral to the centre of the monster using all-purpose household glue.

5 Apply a few glue dots to the fingers of the monster. Bend the fingers over and stick them to the numeral. Press firmly in place.

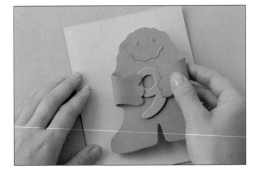

6 Stick the monster to the front of the card using all-purpose household glue.

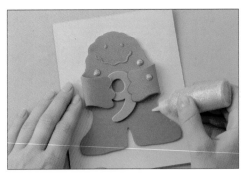

7 Dot the monster with silver glitter relief paint. Leave to dry.

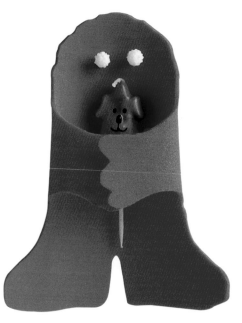

Flying fairy birthday card

Delight a child on their birthday with this pretty fairy bringing good wishes. She is made of scraps of pink silk fabrics and highlighted with lots of glitter paint and tiny star sprinkles.

1 Apply white mulberry paper to white card using spray adhesive. Cut a 29 × 17.5cm/11½ × 7in rectangle of the covered card using a craft knife and metal ruler and working on a cutting mat. Score and fold the card across the centre, parallel with the short edges, using a bone folder.

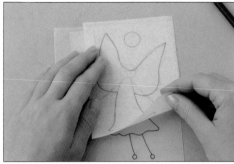

2 Trace the fairy template at the back of the book using a black pen. Tape a piece of iron-on interfacing, shiny side down, to the wrong side of the tracing using masking tape. Trace the dress and wings on to the interfacing with a pencil. Trace the head on another part of the interfacing.

materials and equipment

- white mulberry paper
- white card (stock)
- spray adhesive
- craft knife
- metal ruler
- cutting mat
- bone folder
- tracing paper
- black pen
- pencil
- 12.5 × 10cm/5 × 4in piece of iron-on interfacing
- masking tape
- scraps of bright pink and salmon pink silk dupion fabric
- fabric scissors
- glitter paint in gold and pink
- tiny star sprinkles
- tweezers
- glue dots

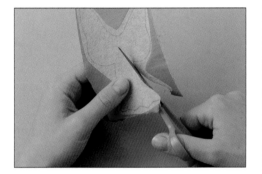

3 Roughly cut out the pieces, leaving a margin all round. Apply the dress and wings to bright pink silk dupion and the head to salmon pink silk dupion by ironing the interfacing, shiny side down, on to the fabric. Cut out the pieces.

4 Centre the dress and wings on the card front, then move it slightly left. Stick to the card front using spray adhesive. Stick the head on top in the same way.

5 Draw the arms, legs, feet, numeral, eyes, mouth and wand on the card lightly in pencil. Create the hair by applying blobs of gold glitter paint around the sides and top of the head. Before the paint dries, carefully place a few tiny star sprinkles on the hair using tweezers. Set aside to dry.

6 Use pink glitter paint to outline the dress, wings and head, then draw the arms, feet and numeral and dot the eyes and mouth. Before the paint dries, carefully place a star sprinkle on each "ankle" using tweezers. Draw the wand with gold glitter paint. Set aside to dry.

7 Stick a few star sprinkles around the numeral, on the dress and on the tips of the wings, using glue dots. Dot gold glitter paint around the number and the stars. Set aside to dry.

For a rugby fan

On this sporty card the ball stands proud of the surface, raised on adhesive foam pads, and the distinctive black lines are applied with relief paint.

materials and equipment

- dark brown, light green and terracotta card (stock)
- craft knife
- metal ruler
- cutting mat
- bone folder
- tracing paper
- pencil
- black relief paint
- cream paper
- glue dots
- adhesive foam pads
- paper glue

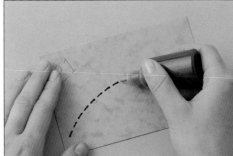

1 Cut a 22 x 15.5cm/8½ x 6¼in rectangle of dark brown card using a craft knife and metal ruler and working on a cutting mat. Score and fold the card across the centre, parallel with the short edges, using a bone folder.

2 Cut a 14.5 x 10cm/5¾ x 4in rectangle of light green mottled card. Use the template at the back of the book to draw the broken line and age number lightly in pencil. Redraw the broken line and number with black relief paint. Leave to dry.

3 Draw the rugby ball on terracotta card. Draw the seam with black relief paint. Draw the lacing panel on cream paper, with the lacing in black relief paint. Leave to dry then cut out the pieces. Stick the lacing panel to the ball using glue dots.

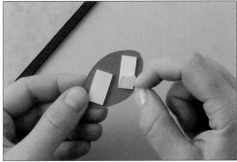

4 Stick the light green rectangle to the card front using spray adhesive. Stick the ball to the light green rectangle using adhesive foam pads. Cut a 21 x 14.5cm/8⅛ x 5¾in rectangle of cream paper for the insert and fold it in half. Run a line of paper glue along the fold and stick the insert inside the card, matching the folds.

Sponged foliage gift wrap

Shapely real leaves, such as the leaves of a Japanese acer tree, are held temporarily in place on a sheet of coloured paper using stencil mount, then the paper is sponged all over with acrylic paint.

materials and equipment

- leaves
- blotting paper
- flower press or heavy book
- stencil mount
- orange paper
- red acrylic paint
- old plate or ceramic tile
- flat paintbrush
- natural sponge
- craft knife

1 Press the leaves between sheets of blotting paper in a flower press or between the pages of a heavy book for a few days. When they are flat, spray the leaves with stencil mount, arrange them on a sheet of orange paper and press in place.

2 Apply some red acrylic paint to an old plate or ceramic tile and spread it out using a flat paintbrush.

Tip
Test your colour choices and choice of stencil on spare paper first.

3 Dab at the paint with a natural sponge.

4 Sponge the paint all over the paper. Leave to dry then peel off the leaves, lifting the edges with the tip of a craft knife blade.

Paisley birthday card

Here, an elegant paisley design is stencilled on an easel-style greetings card, highlighted with silver paint. A panel with a matching stencilled border holds the card upright. The card folds flat to fit a standard envelope.

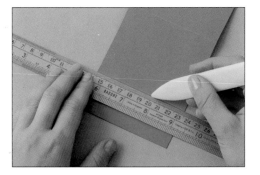

1 Cut a 53 x 9cm/20¾ x 3½in strip of lilac card using a craft knife and metal ruler and working on a cutting mat. Score the card using a bone folder, parallel with the short edges, 4cm/1½in, 11cm/4¼in and 32cm/12½in from one end.

2 Fold the card along the scored lines to form the easel shape. Crease the folds with a bone folder. Open the card out flat again.

materials and equipment

- lilac card (stock)
- craft knife
- metal ruler
- cutting mat
- bone folder
- tracing paper
- pencil
- stencil board
- masking tape
- acrylic paint in jade green and silver
- stencil brush
- kitchen paper

3 Trace the templates of the paisley shape and the border at the back of the book and transfer them to stencil board. Cut out the stencils using a craft knife, working on a cutting mat.

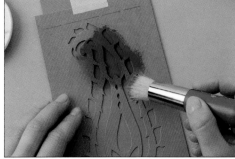

4 Tape the paisley stencil to the front section using masking tape. Pick up a small amount of jade green paint with a stencil brush, dabbing off the excess paint on kitchen paper. Dab the paint through the stencil, holding the brush upright and moving it in a circular motion. Leave to dry.

5 Tape the border stencil to the 4cm/1½in end section and stencil the border with jade green paint. Leave to dry.

6 Stencil the centre and outer edges of the paisley, and the upper edge of the border, with silver acrylic paint. Leave to dry then remove the stencils.

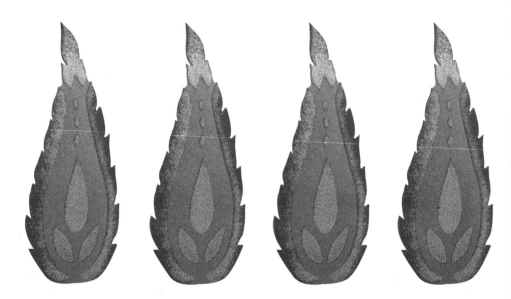

Shell and pearl-trimmed card

This charmingly feminine card in shades of pink combines pearlized paper with a little string of pearls and a delicate shell. It's multilayered for an extra touch of luxury, and would make a perfect birthday greeting for your mother.

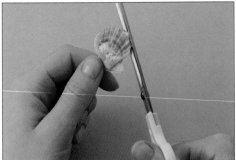

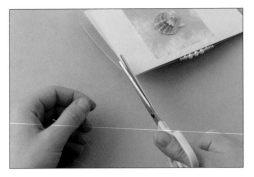

materials and equipment

- 3 x A5 sheets pearlized paper, white, pale pink and pink
- 7cm/2³/₄in square of silver paper
- glue stick
- scallop shell
- adhesive pads
- scissors
- clear bead elastic
- pearl and pink rocaille beads
- craft knife
- metal ruler and cutting mat
- adhesive tape

1 Fold the three sheets of paper in half. Using a craft knife and metal ruler and working on a cutting mat, trim 6mm/¼in from the open side of the pale pink sheet and 1.2cm/½in from the white sheet.

2 Glue the silver paper to the centre front of the white sheet. Stick adhesive pads to the back of the shell, trim any protruding edges, then attach the shell to the centre of the silver paper.

3 Cut a 50cm/20in length of bead elastic and make a double knot close to the centre. Thread on a few pearl beads, alternating them with tiny pink rocailles. Make another knot to secure the beads.

4 Slip the pink papers inside the white paper and make a tiny notch at each end of the fold through all the layers. Tie the elastic so that it lines up the spine of the card and secure with tiny pieces of adhesive tape.

Exotic flowers and lace

The smallest scraps of fabric can be turned into gorgeous cards: combine an exuberant print like these tropical flowers with fragments of old lace.

materials and equipment

- white card (stock)
- craft knife
- metal ruler
- cutting mat
- white card blank
- cream tissue
- glue stick
- double-sided tape
- white lace
- green and white printed paper
- flowery fabric
- fabric scissors
- fusible bonding web
- iron
- scrap paper

1 Cut a piece of white card the same size as the card blank. Cut a piece of tissue 2cm/¾in larger all round and stick the card in the centre using a glue stick. Fold the edges to the back and stick with double-sided tape. Glue a strip of lace to each side and stick the raw edges down at the back.

2 Use double-sided tape to stick a strip of patterned paper to the centre of the card, overlapping the edges of the lace panels.

3 Roughly cut out three flower motifs from the patterned fabric. Iron them on to fusible bonding web, following the manufacturer's instructions, then cut out carefully.

4 Peel off the backing, arrange them attractively on the card and iron in place with a cool iron. Use a sheet of paper to protect the surface of the card when pressing. With double-sided tape, stick the card to the front of the card blank.

Punched gift tag

Experiment with decorative-edged scissors and paper punches in a range of designs to make dramatic gift tags from contrasting card and paper.

materials and equipment

- red card (stock)
- wavy-edged scissors
- ruler
- bone folder
- olive green textured paper
- spiral paper punch about 10 x 8mm/³/₈ x ⁵/₁₆in
- cutting mat
- 4mm/⁵/₃₂in hole punch
- tack hammer
- 3mm/¹/₈in hole punch
- 2mm/¹/₁₆in hole punch
- pale blue paper
- spray adhesive
- double-sided tape

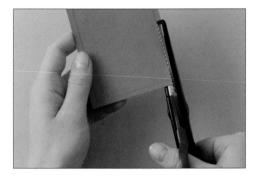

1 Cut a 12 x 7cm/4¾ x 2¾in rectangle of red card with wavy-edged scissors. Score the card parallel with the short edges, 5cm/2in from one side, using a bone folder.

2 Cut a 6 x 4cm/2¼ x 1½in rectangle of olive green textured paper using wavy-edged scissors. Punch a row of four spirals along each long edge using a paper punch.

3 Resting on a cutting mat, punch a 4mm/⁵/₃₂in hole in the centre of the olive green rectangle using a hole punch and a tack hammer. Punch 3mm/¹/₈in and 2mm/¹/₁₆in holes above and below the centre. Stick the rectangle to the front of the tag using spray adhesive.

4 Cut a 6cm/2½in square of pale blue paper using wavy-edged scissors. Resting on a cutting mat, punch a row of five 2mm/¹/₁₆in holes, 5mm/¼in from one edge. Stick the square inside the tag using spray adhesive. Stick a piece of double-sided tape to the back of the tag ready to stick to the gift.

Greek urn card

Make this classically inspired birthday card for a friend who is keen on antiques or travelling. The Greek urn is made of beautifully textured paper.

materials and equipment

- beige paper with embedded fibres and metallic fragments
- cream card (stock)
- spray adhesive
- craft knife
- metal ruler
- cutting mat
- bone folder
- tracing paper
- pencil
- fine black pen
- black paper
- double-sided tape
- 15mm/⅝in leaf sprig paper punch

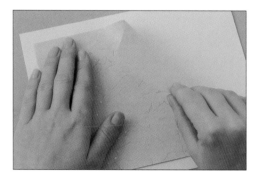

1 Apply beige paper with embedded fibres and metallic fragments to cream card using spray adhesive. Cut a 19 × 14.5cm/7½ × 5¾in rectangle of the covered card using a craft knife and metal ruler and working on a cutting mat. Score and fold the card across the centre, parallel with the short edges, using a bone folder.

2 Use the template from the back of the book to cut the urn from the folded card, aligning the left side with the folded edge. Refer to the template to draw the harp strings on the card using a black pen.

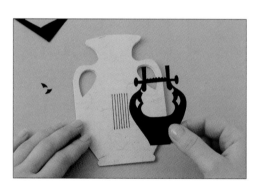

3 Use the template to cut the harp from black paper. Stick the harp to the front of the card using spray adhesive.

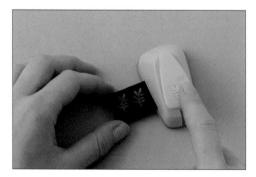

4 Stick double-sided tape to the back of a piece of black paper. Use a 15mm/⅝in leaf sprig paper punch to punch three sprigs. Peel off the backing paper and stick the sprigs in a row around the neck of the urn.

Japanese paper fans

Origami paper is easy to fold sharply, so it is ideal for making tiny fans. It is available in traditional Japanese patterns in rich colours.

materials and equipment

- 3 sheets of patterned origami paper
- craft knife
- metal ruler
- cutting mat
- masking tape
- double-sided tape
- gold paper
- scissors
- 18cm/7in square of dark blue card (stock)
- green textured paper
- 8 x 16cm/3 x 6in rectangle of red mulberry paper
- bone folder
- glue stick
- glue dots
- small gold tassel

1 Cut three 15 x 2.5cm/6 x 1in strips from the different origami papers, using a craft knife and metal ruler and working on a cutting mat. Make narrow accordion folds in each strip of paper to make fan shapes.

2 Hold the end of one pleated paper tightly together and bind it with a thin strip of masking tape. Stick double-sided tape to the back of a small strip of gold paper and wrap this around the masking tape. Trim the ends neatly. Complete the other two fans in the same way.

3 Fold the blue card in half and go over the crease with a bone folder to sharpen it. Tear the edges of the green paper against a ruler so that it is 5mm/¼in smaller all round than the red paper.

4 Glue the red and green papers to the folded card using glue stick, then stick the fans in place with glue dots. Attach a small gold tassel just below the lowest fan.

Card for a keen traveller

The design of this card was inspired by the plane motif on the envelope, and the border is copied from a standard airmail envelope.

materials and equipment

- airmail envelope
- colour photocopier
- scissors
- A4 sheet of dark grey Ingrès paper
- bone folder
- 10 x 16cm/4 x 6½in rectangle of parchment paper
- glue stick
- luggage label
- adhesive foam pads
- airmail sticker
- adhesive tape

1 Colour photocopy the envelope, enlarging it to 200 per cent of the original size. Cut out the red and blue border strips and the plane image.

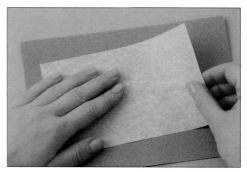

2 Fold the Ingrès paper in half and rub over the crease with a bone folder. Glue the parchment to the centre front of the card using a glue stick, setting it at a slight angle.

3 Glue the red and blue border strips around the edge of the parchment.

4 Stick the plane across the luggage label using a glue stick, then attach the label to the parchment using adhesive foam pads. Add an airmail sticker to one corner of the card and secure the loose ends of the string on the label on the back of the card with a small piece of adhesive tape.

Tip
Photocopy any collection of interesting stamps, postmarks and travel souvenirs to make similar collage cards.

Origami gift box

This handsome box with its well-fitting lid is constructed using a traditional Oriental paper folding technique, though the modern papers used here give it an up-to-the-minute look. The box needs no adhesive to hold its shape and can be used to present a lightweight birthday gift such as a silk scarf, or be filled with shredded tissue to enclose a piece of jewellery.

materials and equipment

- 30cm/12in square sheet of green handmade paper
- 30cm/12in square sheet of pale blue flock gift wrap

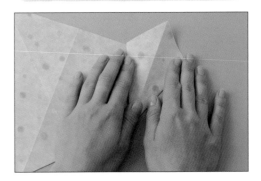

I Refer to the diagrams at the back of the book to fold the squares along the solid and broken lines with the wrong sides of the papers facing. Open the squares out flat again. (The box can be made in any size. The finished container will be a third of the size of the paper. For example, a 30cm/12in square of paper will make this 10cm/4in square box.)

2 Refold the green paper for the box diagonally in half. Fold the corners at the end of the diagonal fold inwards along the broken lines.

3 Stand the corners upright along the broken lines then squash them flat, matching the diagonal fold line to the broken lines. Crease along the new folds. Open the paper out flat and repeat steps 2–3 on the other diagonal fold.

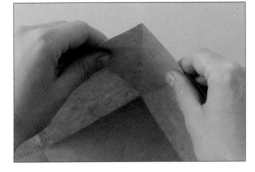

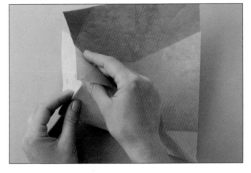

4 With the wrong side facing upwards, fold the green paper along two adjacent base lines to form two sides of the box. Bring the broken lines together, folding the excess paper at the corner inside.

5 Fold the triangle at the top of the corner over the corner to hold it in place. Recrease the corner fold to define it. Repeat to form the remaining corners of the box.

6 To make the lid, with the wrong side of the pale blue flock gift wrap facing upwards, fold two opposite corners to meet at the centre. Fold again so that the broken lines meet at the centre.

7 Lift the edges along the last pair of folds to form two opposite sides of the lid. Lift and fold the third side of the lid by bringing the third broken line level with the upright sides, folding the excess paper inwards at the corners.

8 Repeat with the fourth broken line on the fourth side. Fold down along the broken lines so that the points meet at the centre on the underside of the lid.

Exotic insect birthday card

Silk painting is a very popular pastime. If you are new to the craft, making a panel for a greetings card is a great way to start as it is a fairly small-scale project. Practise on some spare fabric to see how the paints and the outliner work. This fabulous creature is set in a card of marbled paper trimmed with metallic cord and beads.

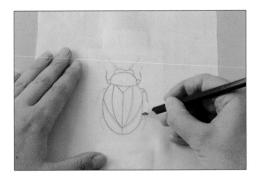

1 Trace the template at the back of the book using a black pen. Tape the silk centrally on top with masking tape. Trace the design lightly on the silk with a pencil. Stretch the silk in an embroidery hoop.

2 Draw the design on the silk with silver silk outliner, starting at the centre of the design and working outwards so that you do not smudge the outliner. Leave to dry. If any lines are very thin or do not join up, go over them again so that the paint colour will not be able to seep through.

3 Dip the paintbrush into the green silk paint and press the brush into the centre of one area. The paint will flow within the outline. Add more paint if necessary until the area is filled. Paint all the areas green or blue. Leave to dry then press the silk between two layers of white tissue paper.

4 Stick the silk to a piece of white paper with spray adhesive, smoothing the silk outwards from the centre. Use a pencil to draw a 10.5 x 8cm/4⅛ x 3⅛in rectangle on the silk with the insect centred inside it. Cut out the rectangle with scissors.

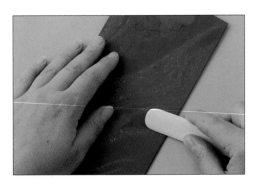

5 Apply marbled paper to blue card with spray adhesive. Cut a 22 x 20cm/8¾ x 8in rectangle of the covered card using a craft knife and metal ruler and working on a cutting mat. Score and fold the card across the centre, parallel with the long edges, using a bone folder. Open the card out flat.

materials and equipment

- tracing paper and pen
- 20cm/8in square of white habotai silk
- masking tape
- pencil
- 15cm/6in embroidery hoop (frame)
- silver silk outliner (gutta)
- green and blue silk paints
- medium artist's paintbrush
- white tissue paper
- white paper
- spray adhesive
- scissors
- iron
- blue and green marbled paper
- blue card
- craft knife
- metal ruler and cutting mat
- bone folder
- 5mm/¼in-wide double-sided tape
- light green paper
- paper glue
- 1.4m/55in silver cord
- 2 blue beads with large holes

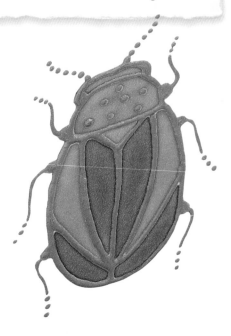

6 For the window, draw an 8.5 × 6cm/3⅜ × 8⅜in rectangle centrally on the card front, 2cm/¾in down from the upper edge. Cut out the rectangle using a craft knife and metal ruler on a cutting mat. Apply double-sided tape around the window on the wrong side. Peel off the backing strips and stick the silk behind the window.

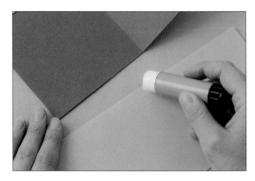

7 Cut a 21 × 18cm/8¼ × 7in rectangle of light green paper for the insert. Fold the paper in half parallel with the short edges. Run a line of glue down the fold using a glue stick and insert the paper in the card, matching the folds.

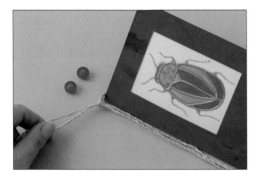

8 Double the silver cord and lay it inside the card with the loop at the top of the spine. Bring the loose ends up the outside of the card and thread them through the loop, pulling the cord tight. Thread a blue bead on to each end of the cord and tie knots to hold the beads in place.

Cupcake gift box

You could make a set of these delightful gift boxes to give your guests at a special birthday tea party. The cake is "iced" with white relief paint.

materials and equipment

- light brown and white card (stock)
- craft knife
- metal ruler
- cutting mat
- bone folder
- double-sided tape
- pair of compasses and pencil
- bright pink corrugated card (stock)
- all-purpose household glue
- tracing paper
- pencil
- masking tape
- white relief paint
- tiny pink star-shaped sprinkles
- bradawl
- 10cm/4in fine green thonging
- 12mm/¹/₂in diameter red bead

1 Cut a 20 x 5cm/8 x 2in strip of light brown card for the side of the box. Score and fold the box side 1.5cm/⅝in from, and parallel with, one long edge for the tabs. Open the strip out flat again. On the wrong side, stick double-sided tape along the tab section and to one end of the strip. Cut a line of V-shaped tabs along the taped edge.

2 With compasses, draw a 5.5cm/2⅛in diameter circle on light brown card for the base and cut it out. Starting at the untaped end, wrap the box side around the base, sticking the tabs under the box. Overlap the ends of the box side and stick together.

3 Cut a 21 x 2.5cm/8¼ x 1in strip of bright pink corrugated card, cutting the short edges parallel with the corrugations. Glue the corrugated strip around the side of the box using all-purpose household glue, matching the lower edges and overlapping the ends of the strip.

4 Cut a 21 x 1.5cm/8¼ x ⅝in strip of light brown card for the lid rim. Apply double-sided tape to one end of the rim on the wrong side. Starting at the untaped end, wrap the lid rim around the box side, resting it on the corrugated card. Do not pull the strip tight, as it needs to slip on and off the box easily. Overlap the ends of the lid rim and stick together.

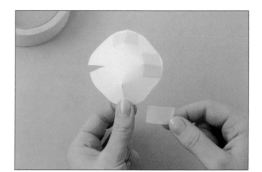

5 Use the template at the back of the book to cut a lid from white card. Bring the straight edges together and join them with strips of masking tape.

6 Run all-purpose household glue along the top of the lid rim. Press the lid centrally on top. Weight the lid while the glue dries.

7 Apply white relief paint liberally to the lid. Sprinkle tiny pink star-shaped sprinkles on the lid. Set aside to dry.

8 Pierce a hole through the centre of the lid using a bradawl. Knot the end of a 10cm/4in length of fine green thonging and thread it up through the hole. Glue the knot in place and dab glue on the thonging just above the hole. Thread on a 12mm/½in diameter red bead. Cut the thonging 2.5cm/1in above the bead.

Eightieth birthday card

Hand embroidery on paper is very effective if the design is simple. A matching tag proudly announces the important birthday that the card celebrates.

materials and equipment

- lilac paper with a silver glitter pattern
- lilac, white and silver card (stock)
- spray adhesive
- cutting mat
- craft knife
- metal ruler
- bone folder
- tracing paper
- pencil
- masking tape
- bradawl
- 4mm/^{5}/$_{32}$in round diamanté sticker
- crewel embroidery needle
- pink and green stranded embroidery thread (floss)
- 4mm/^{5}/$_{32}$in crystal heart sticker
- 4mm/^{5}/$_{32}$in purple brad
- 5mm/^{1}/$_4$in-wide double-sided tape

1 Apply patterned lilac paper to lilac card using spray adhesive. Cut an 18cm/7in square of the covered card using a craft knife and metal ruler on a cutting mat. Score and fold the card across the centre.

2 Cut a 6 × 5cm/2½ × 2in rectangle of white card. Cut a tag from silver card. Trace the templates at the back of the book. Tape the tracings to the cards and pierce holes at the dots using a bradawl.

3 Stick a round diamanté sticker to the centre of the flower. Thread a needle with pink embroidery thread and work lazy daisy stitch around the stone. Stitch the stem with running stitch and the leaves with single stitches, using green thread.

4 On the tag, stitch the number 80 with back stitch using pink embroidery thread. Stick a heart sticker to the tag and attach the tag to the embroidered rectangle with a purple brad through the remaining holes. Stick the rectangle to the card front, 2cm/¾in below the upper edge, using double-sided tape.

Pleated gift wrap

This is an elegant way to wrap a boxed gift for an important birthday. Pearlized gift wrap is neatly pleated around the present, then decorated with a row of jewellery stones in co-ordinating shades.

materials and equipment

- turquoise pearlized gift wrap
- scissors
- pencil
- ruler
- clear adhesive tape
- double-sided tape
- 3 oval jewellery stones in shades of turquoise
- glue dots

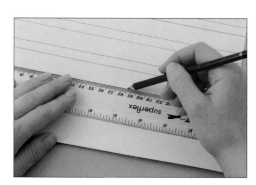

1 Cut the gift wrap large enough to cover the present, adding 8cm/2¼in to the depth to allow for the pleats. On the wrong side of the paper, draw lines across the gift wrap with a pencil and ruler 2cm/¾in then 1cm/⅜in apart three times. Draw a final line 2cm/¾in away from the previous one.

2 Fold the gift wrap along the first line with right sides facing. Match the fold to the second line.

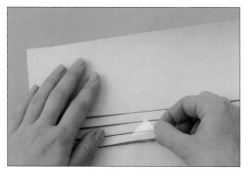

3 Continue folding along the lines and matching the folds to the following lines. Stick two pieces of clear adhesive tape to the folds to hold the pleats in place.

4 Wrap the boxed present, placing the pleats on the top of the box and sticking the gift wrap with double-sided tape. Stick three oval jewellery stones in a row on one of the pleats using glue dots.

Ninetieth birthday card

The shiny number 90 is made from fine embossed metal, with the neat finishing touch of a red star-shaped jewel.

materials and equipment

- gold card (stock)
- craft knife
- metal ruler
- cutting mat
- bone folder
- turquoise translucent paper
- spray adhesive
- tracing paper
- pencil
- paper scissors
- masking tape
- fine brass embossing metal
- old pair of scissors
- bradawl
- kitchen paper
- embossing tool or spent ballpoint pen
- all-purpose household glue
- red star jewellery stone sticker

3 Place the numerals right side down on two sheets of kitchen paper. Use an embossing tool or a spent ballpoint pen to emboss a simple swirling design.

4 Glue the numerals on the front of the card using all-purpose household glue. Stick a star sticker on the nought.

1 Cut a 24 × 16cm/9½ × 6½in rectangle of gold card. Score and fold the card across the centre. Cut a 15 × 11cm/6 × 4¼in rectangle of translucent paper. Stick the rectangle centrally to the front of the card.

2 Use the template at the back of the book to cut the number 90 from tracing paper. Tape the template face down on a piece of fine brass embossing metal. Draw around the numerals with a pencil, pressing firmly to make an indentation. Cut out the numerals using an old pair of scissors. To cut out the centres, make a hole with a bradawl first then cut outwards from the hole.

Hundredth birthday card

Frame a photocopy or a reprint of a family heirloom photograph to commemorate a landmark hundredth birthday.

materials and equipment

- apricot and silver card (stock)
- craft knife and metal ruler
- cutting mat
- bone folder
- spray adhesive
- tracing paper and pencil
- photocopy or reprint of family photograph
- bradawl
- 4 x 12mm/^{1}/$_2$in antique-style brads
- white paper
- double hole punch
- 40cm/16in of 15mm/^{5}/$_8$in-wide grey organza ribbon
- scissors

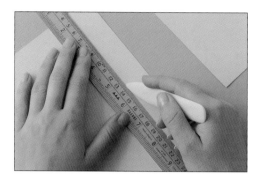

1 Cut two 19 x 18cm/7¾ x 7in rectangles of apricot card using a craft knife and metal ruler and working on a cutting mat. Score and fold one rectangle, which will be the front of the card, 2.5cm/1in in from the long left-hand edge, using a bone folder, to make a hinge. The other rectangle will be the card back.

2 Use the template at the back of the book to cut out the photocopy or reprint, following the broken lines. Cut the frame from silver card following the solid lines. Stick the photograph to the card front then stick the frame on top using spray adhesive.

3 Resting on a cutting mat, pierce holes through the dots on the frame using a bradawl. Insert a brad through each hole. Splay open the prongs inside the card.

4 Cut two 18 x 17cm/7 x 6½in rectangles of white paper for the inserts. Lay the inserts on the back then place the front on top, matching the left-hand edges. Punch a pair of holes centrally in the hinge using a double hole punch. Thread the ribbon through the holes and knot in a double knot. Cut the ribbon ends diagonally.

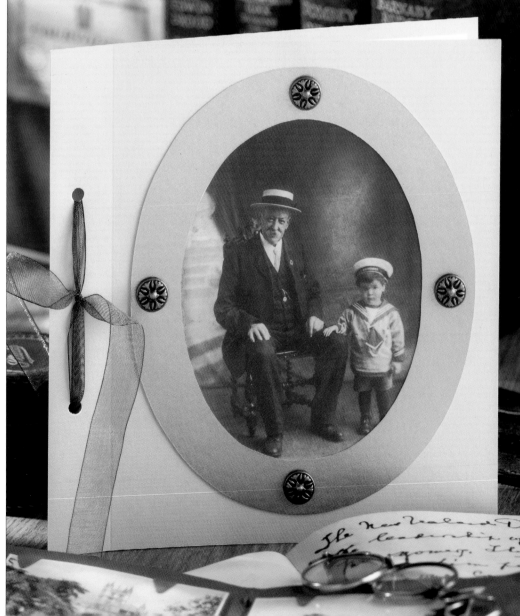

Special occasions

Every landmark occasion deserves to be marked with a greetings card. This chapter presents cards for all of life's red-letter days, from welcoming a new baby to embarking on retirement, and everything in between.

There are witty ideas for the young at heart and elegant sophisticated designs too. There are also some pretty gifts to make and novel ideas for wrapping and trimming presents for christenings, wedding anniversaries and many other occasions.

Birthstone card for a new baby

Birthstones are paired with the months, so choose the relevant stone to welcome a new baby. An amethyst, the birthstone for February, is used here.

materials and equipment

- beige paper with metallic fragments
- beige card (stock)
- spray adhesive
- craft knife
- metal ruler
- cutting mat
- bone folder
- silver textured paper
- cream handmade paper
- polished stone
- bradawl
- 20cm/8in of 0.8mm silver wire
- wire snippers
- double-sided tape

1 Apply beige paper with metallic fragments to beige card using spray adhesive. Cut a 23 × 18cm/9 × 7in rectangle of the covered card. Score and fold the card across the centre, parallel with the short edges, using a bone folder.

2 Against a ruler, tear a 7 × 6cm/3 × 2½in rectangle of silver textured paper and a 6 × 5cm/2½ × 2in rectangle of cream handmade paper. Stick the cream paper in the centre of the silver paper using spray adhesive.

3 Place the paper rectangle on a cutting mat and hold the stone in the centre. With a bradawl, pierce about six holes around the stone to thread the wire through.

4 Insert the wire through one hole and bend 2cm/¾in of the end back on the underside of the silver paper. Wrap the wire over the stone and insert it in another hole. Repeat to secure the stone in place, pulling the wire tight each time. Bend back the end of the wire on the underside of the paper. Snip off the excess wire. Stick the paper to the card front using double-sided tape.

Lace pram baby congratulations card

This delicate card uses scraps from the sewing basket to make a pretty pram motif. Use blue paper, as here, for a boy and pink for a baby girl.

materials and equipment

- beige and light blue handmade paper
- ruler
- bone folder
- tracing paper
- pencil
- fabric scissors
- scrap of cream lace
- spray adhesive
- cutting mat
- bradawl
- beige stranded embroidery thread (floss)
- crewel embroidery needle
- 4mm/⁵/₃₂in cream pearl bead
- 2 × 12mm/¹/₂in mother-of-pearl buttons
- glue dots

1 Tear a 20 × 13cm/8 × 5⅛in rectangle of beige handmade paper by tearing against a ruler. Score and fold the card across the centre, parallel with the short edges, using a bone folder.

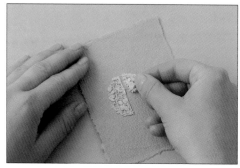

2 Tear a 12 × 9cm/4⅝ × 3½in rectangle of light blue handmade paper by tearing against a ruler. Use the template at the back of the book to cut a pram and hood from a scrap of lace. Stick the lace pieces to the light blue rectangle using spray adhesive.

3 Place the light blue rectangle on a cutting mat. Hold the tracing of the template on top and, using a bradawl, pierce the dots for sewing the handle. Sew along the handle through the holes with a back stitch, using stranded embroidery thread. Catch in a pearl bead with the end stitch and fasten the thread ends at the back of the paper.

4 Glue the light blue rectangle to the card front with spray adhesive. Stick on two buttons for wheels using glue dots.

Naming day gift wallet

Use this colourful wallet to present a christening or naming day gift. The sides of the wallet are joined with pastel-coloured safety pins and the top fastens with ribbon. Make it in pink for a girl or blue for a boy.

materials and equipment

- pink handmade paper
- ruler
- pencil
- cutting mat
- bradawl
- 8 pastel coloured safety pins
- double hole punch
- 50cm/20in of 8mm/⁵⁄₁₆in-wide lilac ribbon
- fabric scissors

1 Carefully tear a 30 × 18cm/12 × 7in rectangle of pink handmade paper by tearing against a ruler. If the paper is strong, dampen it first with a paintbrush, then tear the paper and allow to dry. Fold the paper in half, parallel with the short edges.

2 With a pencil, mark a row of dots 1cm/⅜in in from one side edge, starting 2cm/¾in above the fold. Mark the dots 8mm/⁵⁄₁₆in then 2.5cm/1in apart three times. Mark the top dot 8mm/⁵⁄₁₆in above the previous one. Resting on a cutting mat and using a bradawl, pierce a hole at each dot. Repeat on the other side.

3 Carefully fasten a safety pin through each pair of holes.

4 Punch a pair of holes through both layers centrally at the top of the wallet. Slip the gift inside and fasten with ribbon through the punched holes. Tie the ribbon in a bow and trim the ends diagonally.

Baby's naming day card

By the time a baby's christening or naming celebration takes place, a few months after birth, you may have accumulated a collection of photographs, from which you can make a selection for this pictorial card, converting them to black and white for a timeless look. The dark blue card blank makes a smart change from baby pastels.

materials and equipment

- dark blue card blank
- photograph of baby
- photo corners
- glue stick
- contact or album sheet of small pictures
- scissors and pencil
- 3 swing labels
- bradawl
- 3 brass paper fasteners

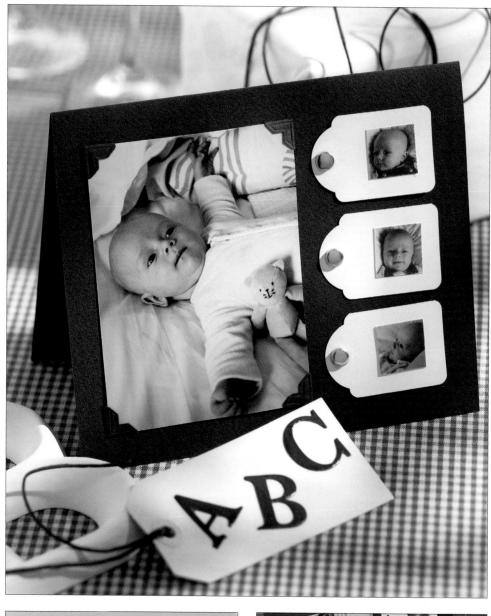

1 Fix the main photograph to the left side of the card front using photo corners. You may need a little extra glue to hold the picture securely in place.

2 Cut out three small images from the contact or album sheet using scissors.

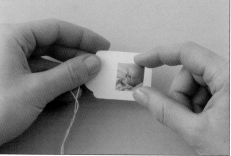

3 Glue the images to the swing labels so that the holes are on the left side. Snip off the ties and arrange the labels on the right-hand side of the card. Mark the positions of the holes.

4 Use the bradawl to pierce the card at each of the three marks. Use brass paper fasteners to fix the tags in place, then glue them down to stop them sliding around.

Confirmation gift box

This beautiful wired flower is used to embellish a ready-made gift box in which to present a small confirmation gift. The beads are threaded on to wire so that they can be bent into petal shapes, and a cluster of twinkling crystal beads forms the centre of the flower. Punch a hole in the box first, if the cardboard shape is stiff.

1 Bend over one end of the wire for 5cm/2in to stop the beads slipping off. Thread on 35 rocaille beads and slip them along to the bend in the wire.

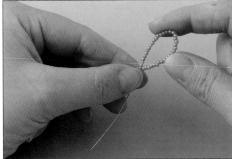

2 Twist the wire around itself a few times at the bend in the wire under the beads, forming the first petal.

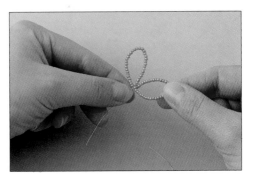

3 Thread on another 35 rocaille beads for the next petal. Twist the wire on itself as before. Make a total of six petals.

4 Bring the long end of the wire up through the centre of the petals and thread on the crystal beads. Coil the beaded wire at the centre of the flower. Pull the end to the underside of the flower and twist the wire on itself a few times to secure.

5 Pierce a hole in the lid of the box where you wish to place the flower. Insert the wire ends through the hole.

6 To neaten the underside of the lid, cut a piece of paper slightly smaller than the lid and stick it in position using spray adhesive. Stick a silver cross sticker inside the box.

materials and equipment

- 55cm/22in of 0.4mm silver wire
- 210 coral coloured pearlized rocaille beads
- 14 transparent 3mm/$\frac{1}{8}$in crystal beads
- wire cutters
- gift box
- bradawl
- co-ordinating paper
- spray adhesive
- silver cross sticker

Handbag invitation

This fun three-dimensional handbag is sure to be cherished. Unfold the bag to write an invitation on the underside, perhaps for a school friends' reunion.

materials and equipment

- pink snakeskin-effect paper
- deep pink card (stock)
- spray adhesive
- tracing paper
- pencil
- craft knife
- metal ruler
- cutting mat
- bone folder
- gold flower paper trim
- paper glue
- gold button
- glue dot

1 Apply pink snakeskin effect paper to deep pink card using spray adhesive. Use the template at the back of the book to cut a handbag from the covered card, using a craft knife and metal ruler and working on a cutting mat.

2 On the wrong side, score the handbag along the broken lines shown on the template, using a bone folder. Fold the handbag with the wrong sides facing, then open the handbag out flat again.

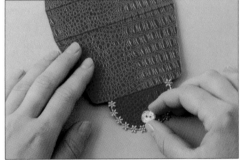

3 On the right side of the bag, stick gold flower paper trim along the curved edge of the flap using paper glue.

4 Stick a glittery gold button to the flap using a glue dot. Fold up the bag, tucking the flap through the handle.

Tip
Use stiff card for the base of this handbag so that it holds its shape once assembled. You could use it to add a small gift as well as an invitation.

Funky foam CD envelope

The soft foam envelope of this design will protect its contents. It would be a great way to wrap a CD gift for a new student starting college. The envelope is designed to be reusable.

materials and equipment

- aquamarine Neoprene foam
- craft knife
- cutting mat
- masking tape
- black A4 sheet of Neoprene foam
- 2 metal eyelets
- eyelet tool
- tack hammer
- bradawl
- 60cm/24in of red plastic thonging
- scissors

1 Cut a flower shape about 5.5cm/2¼in wide from aquamarine Neoprene foam using a craft knife and working on a cutting mat. Use masking tape to hold it in position at the centre of one short edge of a black A4 sheet of Neoprene foam.

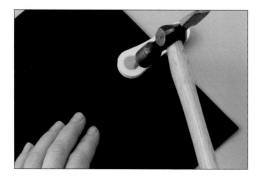

2 Insert a metal eyelet centrally through the flower and wallet and fit it using an eyelet tool and hammer. Remove the masking tape. Insert an eyelet in a matching position at the opposite edge of the sheet.

3 Fold the wallet in half. Resting it on a cutting mat, pierce a row of holes 1cm/⅜in in from each side edge using a bradawl, starting 1cm/⅜in above the fold then piercing the holes 1.5cm/⅝in apart.

4 Knot one end of the plastic thonging and lace it in and out of the holes. Tie a knot at the last hole and cut off the excess. Repeat on the other side. Insert the gift and tie the top with the remaining thonging.

Graduation accordion book

An accordion book has plenty of room to write in and can be filled with messages at graduation time. Beautiful toile de jouy fabric is applied to the cover. The book is decorated with a golden rosette and fastens with ribbon.

materials and equipment

- white card (stock)
- craft knife
- metal ruler
- cutting mat
- toile de jouy fabric
- fabric scissors
- PVA (white) glue
- white paper
- bone folder
- 15cm/6in of 15mm/⁵⁄₈in-wide gold ribbon
- needle and thread
- 12mm/¹⁄₂in diameter gold jewellery stone
- glue dots
- 70cm/28in of 5mm/¹⁄₄in-wide gold ribbon
- hole punch and tack hammer

1 For the back and front of the book, cut two 20 × 10.5cm/8 × 4¼in rectangles of white card using a craft knife and metal ruler and working on a cutting mat.

2 Cut two 23 × 13.5cm/9¼ × 5½in rectangles of fabric. Place the fabric face down and place each cover on top. Turn in the corners of the fabric diagonally and stick them to the card using PVA glue, then fold in the edges and stick them in place.

3 Cut a 60 × 19.5cm/24 × 7¾in rectangle of white paper for the pages. Score and fold the paper at 10cm/4in intervals, parallel with the short edges, using a bone folder.

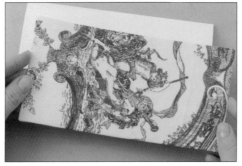

4 Spread PVA glue on one end page and stick one of the covers to it centrally. Stick the second cover on the other end page.

5 Sew a running stitch along one edge of the 15mm/⅝in-wide gold ribbon. Gather the ribbon tightly, overlapping the ends, and fasten off the thread securely.

6 Stick a 12mm/½in diameter gold jewellery stone to the centre of the ribbon rosette with a glue dot.

7 Fold a 12cm/4¾in length of narrow gold ribbon in half. Stick the fold with a glue dot behind the rosette and trim the ends diagonally. Stick the rosette to the front cover of the book with glue dots.

8 Resting on a cutting mat, punch a 3mm/⅛in hole centrally in the right-hand edge of the front cover using a hole punch and tack hammer. Thread the remaining length of narrow gold ribbon through the hole and tie in a bow around the card.

Painted pebbles leaving card

This restful design of striped pebbles looks effective on lovely handmade paper but is simple to paint. Although the card is dark blue, an insert inside the card means that a message can be written on light coloured paper.

materials and equipment

- dark blue card (stock)
- craft knife
- metal ruler
- cutting mat
- bone folder
- mustard yellow handmade paper
- pencil
- medium and fine artist's paintbrushes
- acrylic paints in grey and off-white
- spray adhesive
- mottled yellow paper
- paper glue

3 Draw three pebbles on the mustard paper panel with a pencil. Paint the pebbles with grey acrylic paint using a medium paintbrush. Leave to dry.

4 Using a fine paintbrush, paint thin stripes across the pebbles with off-white paint. Stick the painting to the front of the card with spray adhesive. Cut a 20 × 17cm/7¾ × 6⅝in rectangle of mottled yellow paper and fold the paper in half parallel with the long edges. Run a line of paper glue along the fold and stick the insert inside the card, matching the folds.

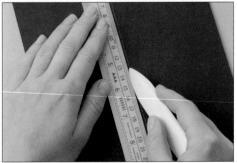

1 Cut a 21 × 18cm/8⅛ × 7in rectangle of dark blue card using a craft knife and metal ruler and working on a cutting mat. Score and fold the card across the centre, parallel with the long edges, using a bone folder.

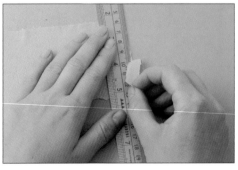

2 Lightly draw a 19.5 × 7.5cm/7½ × 3in rectangle on mustard yellow handmade paper. Moisten the outline with a medium artist's paintbrush and tear the paper along the line against a ruler.

Chinese-stamped gift box

The textured paper cover of this smart gift box gives it an Oriental feel, so it has been decorated with a stamped and embossed bamboo motif. It makes a perfect themed wrapping for a silk scarf.

materials and equipment

- pale grey and black paper
- craft knife
- metal ruler
- cutting mat
- rubber stamp with bamboo design, measuring about 6.5 x 3cm/2½ x 1¼in
- black ink pad
- scrap paper
- embossing powder
- embossing gun or other heat source, such as a toaster
- spray adhesive
- metallic copper and aluminium paper
- white gift box
- 2 chopsticks
- 5mm/¼in double-sided tape
- black imitation suede tape

1 Cut a 7.5 x 4cm/3 x 1¾in rectangle of pale grey paper. Press the rubber stamp on the ink pad. Stamp the motif centrally on the pale grey rectangle.

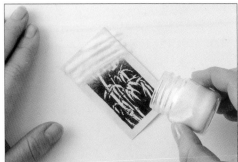

2 Resting the motif on scrap paper, sprinkle embossing powder on the ink. Shake off the excess. Heat the image until the powder melts. Stick the panel on black paper using spray adhesive and trim to leave a 2mm/¹⁄₁₆in border around the grey paper.

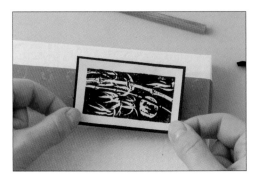

3 Cut a 3cm/1¼in-wide strip of metallic copper and aluminium paper. Position the strip 2cm/¾in in from one edge of the lid and stick the ends of the strip under the lid rim using double-sided tape. Stick the motif to the strip using spray adhesive.

4 Stick two chopsticks to the lid with a few small pieces of 5mm/¼in-wide double-sided tape. Place the present in the gift box. Tie the box with two lengths of black imitation suede tape and stick the ends under the box using double-sided tape.

Slotted butterfly gift box

Give a small engagement present in this pretty gift box. The box can be reused, as the butterfly on the top is in two halves that slot together. Accurate cutting and folding will ensure that the box is symmetrical once assembled. A few diamanté stickers decorate the wings of the butterfly.

1 Trace the template at the back of the book and transfer it to lavender card. Cut out the box using a craft knife and metal ruler, working on a cutting mat.

2 Score the card along the broken lines using a bone folder. Rub away the pencil lines with a pencil eraser. Fold the box along the scored lines then open it out flat again.

materials and equipment

- tracing paper
- pencil
- lavender card (stock)
- craft knife
- metal ruler
- cutting mat
- bone folder
- pencil eraser
- 1.5cm/⅝in-wide double-sided tape
- blue diamanté triangular stickers

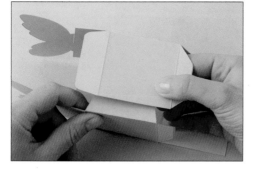

3 Apply double-sided tape to the end and base tabs on the right side. Stick the end tab inside the opposite end of the box.

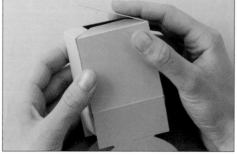

4 Fold in the base tabs. Fold down the base and press the edges firmly on to the tabs to secure it.

5 Tuck the upper side tabs inside the box. Slot the two halves of the butterfly together to fasten the box.

6 Stick blue diamanté triangular stickers on the tips of the wings. To place the stickers accurately, pick them up on the tip of a craft knife blade. Position them using the blade then remove it and press them firmly into place.

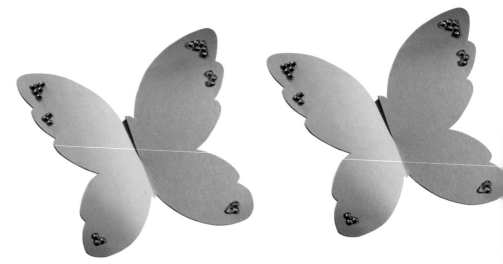

Floral engagement card

Here is a sophisticated and romantic card to celebrate an engagement. A pressed rosebud is wrapped with delicate Japanese paper and set off with lovely paper containing metallic decorations.

materials and equipment

- rosebud
- blotting paper
- flower press or heavy book
- beige paper with metallic fragments
- cream card (stock)
- spray adhesive
- craft knife
- metal ruler
- cutting mat
- PVA (white) glue
- white lightweight Japanese grid paper
- brown paper printed with silver glitter
- bone folder

1 Press a rosebud between sheets of blotting paper in a flower press or within the pages of a heavy book for about 10 days. Apply beige paper with metallic fragments to cream card using spray adhesive. Cut out a 5 × 10cm/2 × 4in rectangle of the covered card using a craft knife and metal ruler and working on a cutting mat. Stick the rosebud centrally to the card using PVA glue.

2 Tear an 8 × 5cm/3¼ × 2in rectangle of Japanese grid paper between your fingers. Wrap the paper diagonally across the stem of the bud, folding the excess to the back of the card. Stick the ends to the back of the card using PVA glue.

3 Apply brown paper printed with silver glitter to cream card using spray adhesive. Cut a 20cm/8in square of the covered card. On the wrong side, score and fold the card across the centre using a bone folder.

4 Stick the wrapped rosebud panel to the card front, 2cm/¾in in from the folded edge and down from the upper edge, using spray adhesive.

Wedding cake card

A special occasion deserves a special card. This exquisite yet elegantly understated card fits the bill when a wedding is being celebrated.

materials and equipment

- white ridged card (stock)
- craft knife
- metal ruler
- cutting mat
- bone folder
- pale yellow paper embedded with metal fragments
- spray adhesive
- gold glitter paint
- fine artist's paintbrush
- 6 tiny diamantés
- tweezers
- 10cm/4in of fine gold cord
- embroidery scissors
- glue dots
- tiny gold heart-shaped sticker
- masking tape
- 3mm/¹/₈in hole punch
- tack hammer
- 45cm/18in length of 3mm/¹/₈in-wide gold ribbon

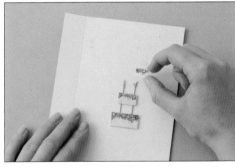

3 Cut the fine cord in half. Arrange the cords 1.2cm/½in apart in the centre of the card front to form the cake columns. Stick the ends in place with glue dots. Stick the tiers on top with glue dots. Stick a gold heart-shaped sticker above the top tier.

4 Tape the front of the card to the back with masking tape, matching the left-hand edges. Resting on a cutting mat, punch a hole at the centre of the hinge using a 3mm/¹/₈in hole punch and tack hammer, then punch two more holes 4.5cm/1¾in above and below the centre. Cut the gold ribbon into three. Thread each length through a hole and tie in a knot around the hinge. Remove the masking tape.

1 Cut two 18 x 13.5cm/7 x 5¼in rectangles of ridged white card using a craft knife and metal ruler and working on a cutting mat. Score and fold one rectangle 2.5cm/1in in from the long left edge to make a hinge. This will be the front of the card. Apply pale yellow paper embedded with metal fragments to a small piece of white card using spray adhesive.

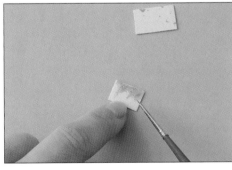

2 Cut one 1.5 x 1cm/⅝ x ⅜in, one 2 x 1.2cm/¾ x ½in and one 3 x 1.5cm/1¼ x ⅝in rectangle of the covered card for the cake tiers. Apply gold glitter paint along the upper edge of the smallest tier. Use a paintbrush to brush the glitter downwards. Place a tiny diamanté on the glitter using tweezers. Repeat to decorate the other tiers. Set aside to dry.

Silver wedding anniversary card

Handmade paper embedded with delicate petals creates a pretty background for pressed larkspur flowers, which are applied to silvered squares on this lovely silver wedding anniversary card.

1 Press three flowers between sheets of blotting paper in a flower press or within the pages of a heavy book. Set aside for about 10 days.

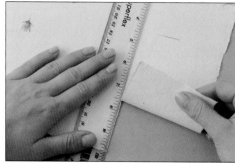

2 Tear a 20cm/8in square of handmade paper by tearing against a ruler. Score and fold the paper across the centre, parallel with the short edges, using a bone folder.

3 Lightly draw a line of three 3cm/1¼in squares down the card front with a pencil, starting 3cm/1¼in down from the upper edge and spacing them 2cm/¾in apart. With a flat paintbrush, apply gold size within the outlines. Set aside for 15 minutes until the size has become tacky.

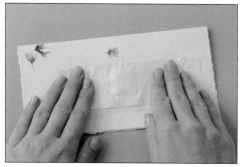

4 Cut a piece of aluminium transfer leaf slightly larger than the group of squares. Lay the aluminium leaf face down on the squares and gently press in place.

5 Peel off the backing paper. Sweep away the excess aluminium leaf using a soft brush. If the aluminium leaf has not adhered to the size in places, press a piece of the left-over leaf in place.

6 Stick a pressed flower over the lower left corner of each square with PVA glue, positioning the flowers with tweezers.

materials and equipment

- 3 larkspur flowers
- blotting paper
- flower press or heavy book
- handmade paper with embedded petals
- ruler
- bone folder
- pencil
- flat paintbrush
- 15-minute gold size
- scissors
- sheet of aluminium Dutch metal transfer leaf
- soft brush
- PVA (white) glue
- tweezers

Tip
Pressed flowers, in perfect condition, can be purchased from craft supply stores if time is short.

Flowers can also be pressed using a microwave, which speeds up the drying process. Use tweezers to handle small, delicate flowers in order to avoid damaging the petals.

Golden wedding card

Against a background of cream and gold paper, a frayed silk square displays a golden metal heart-shaped plaque. Tiny jewels are a delicate finishing touch.

materials and equipment

- turquoise pearlized card (stock)
- craft knife
- metal ruler
- cutting mat
- bone folder
- cream and gold printed paper
- medium artist's paintbrush
- spray adhesive
- scrap of turquoise silk dupion
- fabric scissors
- 2.5cm/1in-wide double-sided tape
- gold heart plaque
- glue dots
- 3 tiny light blue jewellery stone heart stickers

1 Cut a 23 × 16cm/9 × 6¼in rectangle of turquoise pearlized card using a craft knife and metal ruler and working on a cutting mat. Score and fold the card across the centre, parallel with the short edges, using a bone folder.

2 Moisten the outline of a 13.5 × 9cm/5¼ × 3½in rectangle on a piece of cream and gold printed paper using a medium artist's paintbrush. Tear the paper along the moistened lines. Stick the paper centrally to the card front using spray adhesive.

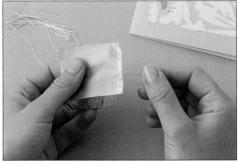

3 Cut a 5.5cm/2¼in square of turquoise silk dupion, cutting the edges along the grain of the fabric. Fray the edges for 1cm/⅜in. Stick a 3cm/1¼in length of 2.5cm/1in-wide double-sided tape to the back of the silk. Stick the silk to the front of the card 2.5cm/1in below the upper edge.

4 Stick a gold heart plaque to the silk square with glue dots. Stick three tiny light blue jewellery stone stickers in a row below the silk square, using the tip of a craft knife blade to lift and position the stickers.

Housewarming in style card

A trio of 50s-style cocktail glasses suggest fun times ahead on this housewarming card. The glasses are cut from translucent papers and accompanied by colourful metallic plastic discs.

materials and equipment

- lime green card (stock)
- craft knife
- metal ruler
- cutting mat
- bone folder
- 5 assorted blue, silver and pink 12mm/¹⁄₂in diameter metallic plastic circles
- glue dots
- tracing paper
- pencil
- turquoise, pink and mauve translucent paper
- white scrap paper
- spray adhesive

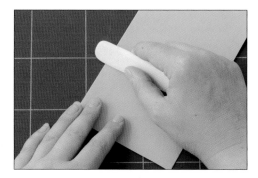

1 Cut a 20 × 18cm/8 × 7in rectangle of lime green card. Score and fold the card across the centre, parallel with the short edges, using a bone folder.

2 Stick five assorted blue, silver and pink metallic plastic circles along the left edge of the card front using glue dots.

3 Trace the cocktail glass templates at the back of the book and transfer to turquoise, pink and mauve translucent paper. Resting on white scrap paper (so that you can see the outlines clearly) on a cutting mat, cut out the glasses using a craft knife.

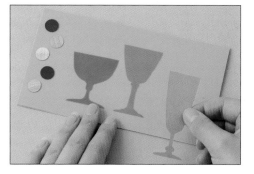

4 Arrange the glasses on the card front. Stick them in place using spray adhesive.

Embossed fish retirement card

You could send this graphic card to a keen angler to celebrate their retirement. The bold fish is fun to make from fine embossing metal.

materials and equipment

- white corrugated card (stock)
- craft knife and metal ruler
- cutting mat
- bone folder
- tracing paper and pencil
- masking tape
- aluminium embossing metal
- embossing tool or spent ballpoint pen
- kitchen paper
- old pair of scissors
- blue spotted paper
- all-purpose household glue
- spray adhesive

1 Cut a 17 x 15cm/6⅝ x 6in rectangle of white corrugated card, with the short edges parallel with the corrugations, using a craft knife and metal ruler and working on a cutting mat. Score and fold the card across the centre, parallel with the short edges, using a bone folder.

2 Trace the fish template at the back of the book and lightly tape the tracing face down on a piece of fine aluminium embossing metal. Draw the details and around the outline with a sharp pencil, using the tracing lines, pressing firmly to make an indentation. Remove the template.

3 Cut a 5.5cm/2¼in square of metal using old scissors. Place the square and the fish, right side down, on two sheets of kitchen paper. Use an embossing tool or ballpoint pen to emboss the details on the fish and dot the square at random. Cut out the fish.

4 Cut a 12 x 5.5cm/4¾ x 2¼in rectangle of blue spotted paper. Arrange the metal square on the card front, overlapped by the blue paper, 5mm/¼in in from the edges of the card front. Stick the metal square with all-purpose household glue and the spotted paper with spray adhesive. Stick the fish on the spotted paper with all-purpose glue.

Watercolour stamped retirement card

Use a rubber stamp with a peaceful design to make this retirement card. The motif is printed with watercolour stamp paints, which can be subtly blended together on the stamp before it is applied to the paper.

materials and equipment

- watercolour paper
- ruler
- medium artist's paintbrush
- bone folder
- 7.5cm/3in square rubber stamp of a peaceful scene
- purple and green watercolour stamp paints
- scrap paper

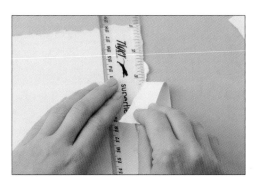

1 Tear a 22 × 14cm/8⅞ × 5½in rectangle of watercolour paper, first drawing a moistened paintbrush along the intended tear lines to weaken them, then tearing against a ruler. Score and fold the card 10.5cm/4¼in from one short edge, parallel with the short edges, using a bone folder.

2 Paint the stamp with purple and green watercolour stamp paints, using a medium artist's paintbrush and blending the colours together on the stamp.

3 Stamp the image on the front of the card, which is the narrower section. Make a firm impression. Leave to dry.

4 Open the card. Resting on scrap paper, moisten the right edge of the lower section. Pick up some green watercolour paint on a moistened paintbrush and run it along the wet edge. Set aside to dry.

Heartfelt sentiments

Sometimes it's good just to send a card to say "hello" and catch up with friends and relatives or convey snippets of news. Telephones and emails have largely superseded the sending of handwritten notes, but it is worth taking the trouble to remedy that, as receiving them is always a joy.

This chapter has cards to make for special friends, to say good luck or best wishes, to wish a loved one good health and send condolences for life's sad occasions too.

Stamped gift wrap

It is simple and economical to make your own gift wrap. This baroque repeat design is stamped in silver on coloured tissue paper. For this kind of repeating design, choose a rubber stamp with a regular shape such as a square or rectangle, so that it is easy to align the motifs accurately to produce an even all-over pattern.

materials and equipment

- mauve tissue paper
- scissors
- scrap paper
- 6cm/2¹/₂in square block design rubber stamp
- silver ink pad

1 If necessary, cut a straight line across the tissue paper with a pair of scissors to make a straight edge. Place the tissue paper on a sheet of scrap paper to protect the work surface from any ink that seeps through the paper. Press the rubber stamp firmly on to the ink pad. Press it a few times to get an even coverage if the stamp is larger than the ink pad.

2 Stamp the paper, matching the upper edge of the stamp to the cut edge. Remove the stamp by lifting it straight up.

3 Press the stamp on to the ink pad again. Stamp the paper, matching the upper edge as before and lining up the left-hand edge with the first stamped image. Continue stamping a row of images.

4 Stamp more rows in the same way to cover the paper completely, lining up the upper edge of the stamp with the lower edge of the motif in the previous row.

Flower garland gift trim

This beautiful gift wrapping idea is very simple to achieve and ideal for a thank you gift. The smaller flowers form a spray of silk delphiniums, threaded on to fine ribbon to form a garland across the top of a wrapped gift.

materials and equipment

- spray of silk
  delphinium flowers
- 3mm/¹/₈in-wide blue ribbon
- 2 x 6mm/¹/₄in lilac beads
- large-eyed needle
- fabric scissors

1 Dismantle the silk delphiniums and select the flowers that suit the size of the wrapped present. Set aside the larger flowers for other craft projects.

2 Thread about four flowers on to a length of narrow blue ribbon long enough to tie around the gift.

3 Bind the ribbon around the wrapped gift and tie the ribbon ends together in a bow. Adjust the threaded flowers so that they lie in a line on the front of the gift.

4 Thread one small flower on to each end of the ribbon. Thread one ribbon end on to a large-eyed needle and thread on a 5mm/¹/₄in bead. Knot the ribbon under the bead and trim the end below the knot. Repeat on the other ribbon end.

Pressed flowers friendship card

Assemble pressed flowers, leaves and scraps of sequin strings and ribbon to make a charming card for a special friend. Pressed flowers are available from craft suppliers if you do not have time to press your own: dyed pressed flowers will add extra splashes of vibrant colour to this eclectic design.

materials and equipment

- assorted flowers and small leaves
- blotting paper
- flower press or heavy book
- light blue A5 handmade paper
- bone folder
- sequin strings in assorted colours
- 15mm/⅝in-wide pink organza ribbon
- fabric scissors
- PVA (white) glue
- cocktail stick (toothpick)
- ruler
- double-sided tape

1 Press a selection of flowers and leaves between sheets of blotting paper in a flower press or within the pages of a heavy book. Set aside for about 10 days. Once pressed, choose the best colours for your card, and discard any that have discoloured or are in poor condition.

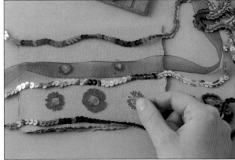

2 Fold an A5 sheet of light blue handmade paper in half, parallel with the short edges, and rub over the fold with a bone folder. Arrange strings of sequins, a length of 15mm/⅝in-wide pink organza ribbon and the pressed flowers and leaves in bands across the front of the card.

3 To stop the sequin strings unravelling, cut the strings approximately 4cm/1½in longer than the width of the card front. Pull a few sequins off each end. Stick the string under the last few sequins using PVA glue, applied with a cocktail stick.

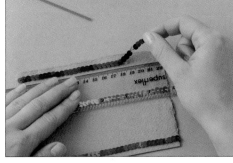

4 Lay a ruler across the card front as a guide to help you stick each sequin string in a straight line. Stick the sequin strings to the card front using PVA glue.

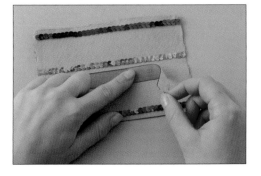

5 Cut the organza ribbon 2cm/¾in longer than the width of the card front. Stick double-sided tape to the ribbon. Peel off the backing strip and stick the ribbon across the card front.

6 Cut the ribbon ends level with the card edges using scissors.

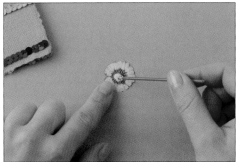

7 Stick the flowers and leaves to the card using PVA glue applied with a cocktail stick.

3D dragonfly card

This richly coloured sparkling dragonfly looks quite intricate, but the technique is simple: the wings are cut around then lifted a little to show the contrasting coloured card underneath. The shimmering effects are created with glitter paint and sequin dust. It's perfect to send to keep in touch with friends.

materials and equipment

- purple and jade green card (stock)
- craft knife
- metal ruler
- cutting mat
- bone folder
- tracing paper
- pencil
- purple glitter paint
- medium and fine artist's paintbrushes
- green, purple and blue sequin dust
- four white photo corners

1 Cut a 32 × 13.5cm/12½ × 5½in rectangle of purple card using a craft knife and metal ruler and working on a cutting mat. Score and fold the card across the centre, parallel with the short edges, using a bone folder.

2 Cut a 13 × 10.5cm/5 × 4in rectangle of jade green card. Use the template at the back of the book to transfer the dragonfly to the rectangle. Apply purple glitter paint to the wings, spreading the glitter paint sparingly with a medium artist's paintbrush. Set aside to dry.

3 Paint the body with purple glitter paint. Use a fine artist's paintbrush to draw out the antennae.

4 Sprinkle green, purple and blue sequin dust on the body before the paint dries. Set the dragonfly aside to dry then shake off the excess sequin dust.

5 Resting on a cutting mat, cut around the edges of the wings using a craft knife, leaving the wings attached at the body. Gently lift the wing tips upwards.

6 Paint the photo corners with purple glitter paint. Sprinkle green, purple and blue sequin dust on the photo corners before the paint dries. Set aside to dry.

7 Slip each photo corner on a corner of the jade green rectangle. Moisten the back of the photo corners and press the rectangle centrally to the card front.

Starburst congratulations card

Embossing paper is known today as parchment craft. This fabulous starburst is embossed through a stencil and the raised surface is coloured with silver metallic wax. The vibrant colours are a great choice for a congratulations card.

2 Turn the paper over. Rub through the stencil cut-outs with a small ball embossing tool or the handle end of a fine artist's paintbrush. The design will be embossed on the right side.

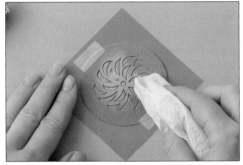

3 Turn the paper over leaving the stencil attached. To highlight the embossing, rub silver metallic wax sparingly on the starburst using kitchen paper. Draw around the stencil. Remove the stencil and cut out the circle. Cut a 7.5cm/3in diameter circle of silver card.

4 Cut a 21 x 13.5cm/8¼ x 5¼in rectangle of orange card using a craft knife and metal ruler and working on a cutting mat. Score and fold the card across the centre, parallel with the short edges, using a bone folder. Using spray adhesive, stick the embossed circle to the circle of silver card, then stick the panel to the card front.

materials and equipment

- tracing paper and pencil
- stencil board
- cutting mat
- craft knife
- masking tape
- pink paper
- small ball embossing tool or fine artist's paintbrush
- silver metallic wax
- kitchen paper
- silver and orange card (stock)
- metal ruler
- bone folder
- spray adhesive

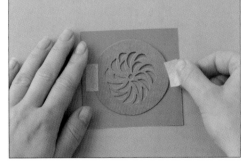

1 Trace the template at the back of the book and transfer the design to stencil board. Resting on a cutting mat, cut out the starburst and circle using a craft knife. Tape the stencil to the right side of a piece of pink paper using masking tape.

Tip

Clashing pink and orange perfectly complement the spiral motif of the stencil. Use colours with a similar tonal value.

Good luck wallet

This pretty wallet decorated with cutwork clover leaves is ideal to contain a card or small gift to wish someone good luck on an important day.

materials and equipment

- green paper
- craft knife
- metal ruler
- cutting mat
- pencil
- tracing paper
- 50cm/20in of 2.5cm/1in-wide aquamarine ribbon
- 5mm/¼in wide double-sided tape
- embroidery scissors

1 Cut out a 22 x 16.5cm/8⅝x 6½in rectangle of green paper using a craft knife and metal ruler and working on a cutting mat. Fold the paper in half parallel with the short edges.

2 Draw four lines 3cm/1¼in long across the centre of the wallet, parallel with the fold and 2.5cm/1in, 4.5cm/1¾in, 7cm/2¾in and 9cm/3¼in above it. Resting the folded paper on a cutting mat, cut along the lines through both layers using a craft knife and a metal ruler, to make four slits.

3 Open the card out flat. Use the template at the back of the book to draw a four-leaf clover on the front of the wallet, 2cm/¾in from one side edge. Turn the template over and repeat on the opposite side. Resting on a cutting mat, cut out the designs with a craft knife.

4 Thread ribbon in and out of the slits. Apply 5mm/¼in-wide double-sided tape to the side edges on the wrong side of the front and stick the front to the back. Slip the card or gift inside and fasten the ribbon. Trim the ends of the ribbon diagonally.

Tip
Punch motifs along the side edges of the wallet using a paper punch if you prefer.

Silvered thank you notelets

Masses of images are readily available in copyright-free books for craftwork. These découpage notelets combine copyright-free vintage motifs with silvered circles. They are quick to make, so a whole series would be easy to produce.

materials and equipment

- lilac card (stock)
- craft knife
- metal ruler
- cutting mat
- bone folder
- pair of compasses
- pencil
- flat paintbrush
- 15-minute gold size
- scissors
- aluminium Dutch metal transfer leaf
- soft brush
- image from copyright-free book
- spray adhesive

1 Cut a 20 × 15cm/8 × 6in rectangle of lilac card using a craft knife and metal ruler and working on a cutting mat. Score and fold the card across the centre, parallel with the short edges, using a bone folder.

2 With the fold horizontal, lightly draw a 4.5cm/1¾in diameter circle on the card front using a pair of compasses and a pencil, 2.5cm/1in in from the fold and left edge. Using a flat paintbrush, apply gold size to the circle. Set the card aside for 15 minutes until the size has become tacky.

3 Cut a piece of aluminium transfer leaf slightly larger than the circle. Lay the aluminium leaf face down on the circle and gently press in place, then peel off the backing paper.

4 Sweep away the excess aluminium leaf using a soft brush. If the aluminium leaf has not adhered in places, press pieces of the left-over leaf into the gaps.

5 Select an image from a copyright-free book and cut it out with a craft knife, resting on a cutting mat.

6 Stick the image to the card front using spray adhesive, overlapping the circle.

Pin-pricked letterhead

Writing paper with an unusual decorative heading is often expensive to buy, but you can create your own stationery for special letters using this easy pin-pricking technique. Experiment with the design: the pricked motif can be positioned at a corner, if you prefer. This would make a perfect letter in which to accept an invitation.

1 Trace the template at the back of the book with a pencil. Tape the tracing right side down to the top of a sheet of writing paper using masking tape. Draw over the outline to transfer it.

2 Remove the tracing. Resting the writing paper on a cutting mat, cut the upper edge with a craft knife.

materials and equipment

- tracing paper
- pencil
- A5 sheets of beige writing paper
- masking tape
- cutting mat
- craft knife
- 2mm/$^{1}/_{16}$in hole punch
- tack hammer
- bradawl

3 Tape the tracing right side up on the writing paper using masking tape. Resting on a cutting mat, punch a hole at each large dot using a 2mm/$^{1}/_{16}$in hole punch and a tack hammer.

4 Pierce holes at the small dots using a bradawl. Remove the tracing. If you wish, carefully enlarge some of the holes with the bradawl. You could cut the flap of an envelope to match the letterhead.

Musical instrument stickers card

Outline stickers are finely detailed adhesive motifs that are available in lots of themes, and they are a very useful way to personalize your greetings cards quickly and easily. The electric musical instrument stickers used here would be ideal decorations for a greetings card to send to a music-mad friend.

materials and equipment

- bright green card (stock)
- craft knife
- metal ruler
- cutting mat
- bone folder
- lime green paper
- spray adhesive
- musical instrument
  outline stickers

1 Cut a 24 × 17cm/9½ × 6½in rectangle of bright green card using a craft knife and metal ruler and working on a cutting mat. Score and fold the card across the centre parallel with the short edges using a bone folder.

2 Cut out triangles from lime green paper. Arrange the triangles on the front of the card and stick in place using spray adhesive.

3 Peel the instrument stickers from the backing paper and apply to the card front.

4 As a finishing touch, stick a few music note stickers inside the card.

Retro photo and cross-stitch card

Remind a friend of happy times by using a favourite old photograph to make this charming card. A '50s-style printed paper was chosen for the background as it suits the period of the photo. Photocopy an original photograph or scan and reprint it using a computer, and adjust the size and shape of the card to suit your print.

1 Apply printed paper to pale yellow card using spray adhesive. Allow to dry. Cut a 28.5 × 21cm/11¼ × 8¼in rectangle of the covered card using a craft knife and metal ruler and working on a cutting mat. Score and fold the card 13.5cm/5¼in from, and parallel with, one short edge, using a bone folder.

2 Cut out the copy of the photograph and a 17 × 10.5cm/6¾ × 4in rectangle of pale yellow card using a craft knife and metal ruler and working on a cutting mat. Stick the image in the centre of the card rectangle using spray adhesive.

materials and equipment

- 1950s-style printed paper
- pale yellow card (stock)
- spray adhesive
- craft knife
- metal ruler
- cutting mat
- bone folder
- **photocopy or reprint of photograph**
- bradawl
- crewel embroidery needle
- **green stranded embroidery thread (floss)**

3 Resting on a cutting mat, use a bradawl to pierce two rows of holes to make three cross stitches 1cm/⅜in in from the top and bottom edges of the photograph.

4 Thread a crewel embroidery needle with green stranded embroidery thread. Sew cross stitches at the pierced holes, knotting the thread on the underside of the card to start and finish.

5 Stick the pale yellow rectangle centrally to the card front using spray adhesive.

6 Tear a strip of the printed paper approximately 8mm/⅜in wide. Stick the strip to the inside back of the card using spray adhesive, aligning the straight edge with the right edge of the card. Trim the ends level with the card.

Flowery get well card

This pretty design, with its fresh, springlike colours and simple flowers, makes a great card to send to a friend to cheer them up and wish them well. It is very simple to make, as the flowers are punched from glittery papers in assorted colours with a flower-shaped punch and then dotted with glitter paint.

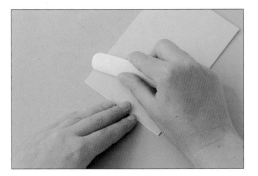

1 Cut a 20 × 15cm/8 × 6in rectangle of pearlized yellow card using a craft knife and metal ruler and working on a cutting mat. Score and fold the card across the centre, parallel with the short edges, using a bone folder.

2 Tear a 10 × 5cm/4 × 2in rectangle of Japanese grid paper by tearing the paper against a ruler. Stick the paper centrally to the card front using spray adhesive.

3 Punch a total of eight flowers from turquoise, lilac and pink glitter papers, using a flower-shaped punch.

4 Apply a 5mm/¼in foam pad to the back of each flower.

5 Peel the backing papers off the foam pads and stick the flowers along the long edges of the Japanese paper.

6 Dot the flower centres with silver glitter paint. Set aside to dry.

materials and equipment

- pearlized yellow card (stock)
- craft knife
- metal ruler
- cutting mat
- bone folder
- **white lightweight Japanese grid paper**
- spray adhesive
- turquoise, lilac and pink glitter paper
- 15mm/⅝in flower-shaped paper punch
- 5mm/¼in adhesive foam pads
- silver glitter paint

Olive bon voyage card

Quilling is the traditional craft of creating pictures with coiled paper strips. This simple motif of a sprig of olives is a great introduction to the craft and makes a charming card to send to someone who is leaving on a trip.

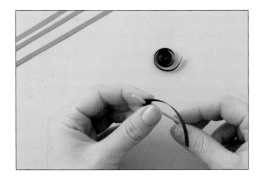

1 Using a craft knife and metal ruler and working on a cutting mat, cut two strips of black paper and three of green paper, each 4mm/³⁄₁₆in wide and 20cm/8in long. Coil the black paper strips and two of the green paper strips tightly around a cocktail stick. Release the coils so that they spring open.

2 Use a cocktail stick to apply PVA glue to the inner surface of one black strip at the outer end of the coil. Stick the end against one side of the coil to form a circle. Repeat with the other coils.

materials and equipment

- craft knife
- metal ruler
- cutting mat
- black and green paper
- cocktail stick (toothpick)
- PVA (white) glue
- A5 beige handmade paper
- bone folder

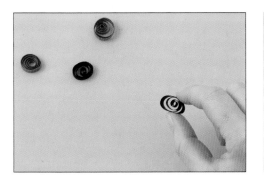

3 Gently squeeze the black circles to form ovals for the olives.

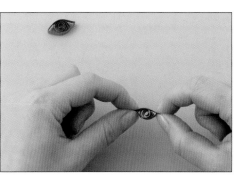

4 Squeeze and pinch the green circles to a point at each side to form the leaves.

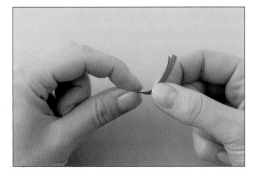

5 Cut the remaining green strip in half. Fold each part in half and hold the fold between a thumb and finger. Starting at the fold, pull the strips between your other thumb and a finger to curl them for the stems.

6 Glue the end of the outer curve 3mm/⅛in from the end of the inner curve.

7 Fold an A5 sheet of beige handmade paper in half, parallel with the short edges, and rub over the fold with a bone folder. Arrange the olives, leaves and stems on the card front. Spread glue on the underside of each coil with a cocktail stick and stick the pieces to the card front.

Accordion leaf card

An accordion or concertina card, sometimes known as a "leporello", allows you to write a long letter within the card. The subdued colours of this elegant version are suitable for a letter of apology or sympathy. The card fastens with ribbon, which can be loosened to allow it to stand upright for display.

materials and equipment

- thick grey paper
- craft knife
- metal ruler
- cutting mat
- bone folder
- 3mm/¹⁄₈in hole punch
- tack hammer
- tracing paper
- pencil
- dark blue, bright olive green and jade green paper
- masking tape
- spray adhesive
- 60cm/24in of 3mm/¹⁄₈in-wide jade green ribbon

1 Cut a 36 × 18cm/14 × 7in rectangle of thick grey paper using a craft knife and metal ruler and working on a cutting mat. Score the card at 9cm/3½in intervals, parallel with the short edges, using a bone folder.

2 Fold the card in accordion folds along the scored lines using a bone folder.

3 Resting on a cutting mat, punch a hole on each side through all the layers, 8cm/3⅛in from the lower edge and 1cm/⅜in inside the folds, using a 3mm/¹⁄₈in hole punch and a tack hammer.

4 Trace the template at the back of the book and transfer the outline of the outer leaf to dark blue, the leaf to bright olive green and the leaf details to jade green paper. Cut them out using a craft knife on a cutting mat.

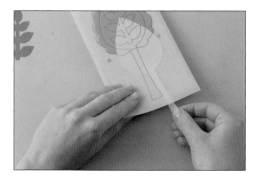

5 To judge the position of the motif on the card, tape the traced template to the top of the card front. Slip the pieces underneath, matching their positions, and stick in place with spray adhesive.

6 Thread the punched holes with ribbon and fasten in a bow on the card front.

TEMPLATES

Enlarge the templates on a photocopier, or trace the design and draw a grid of evenly spaced squares over your tracing. Draw a larger grid on to another piece of paper and copy the outline square by square. Draw over the lines to make sure they are continuous.

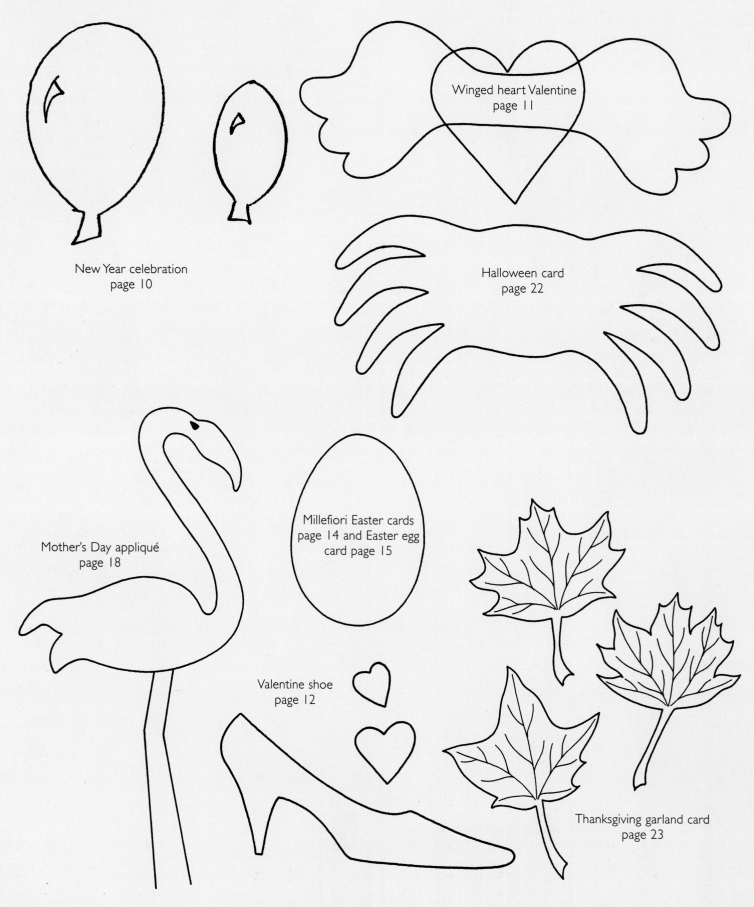

New Year celebration
page 10

Winged heart Valentine
page 11

Halloween card
page 22

Mother's Day appliqué
page 18

Millefiori Easter cards
page 14 and Easter egg
card page 15

Valentine shoe
page 12

Thanksgiving garland card
page 23

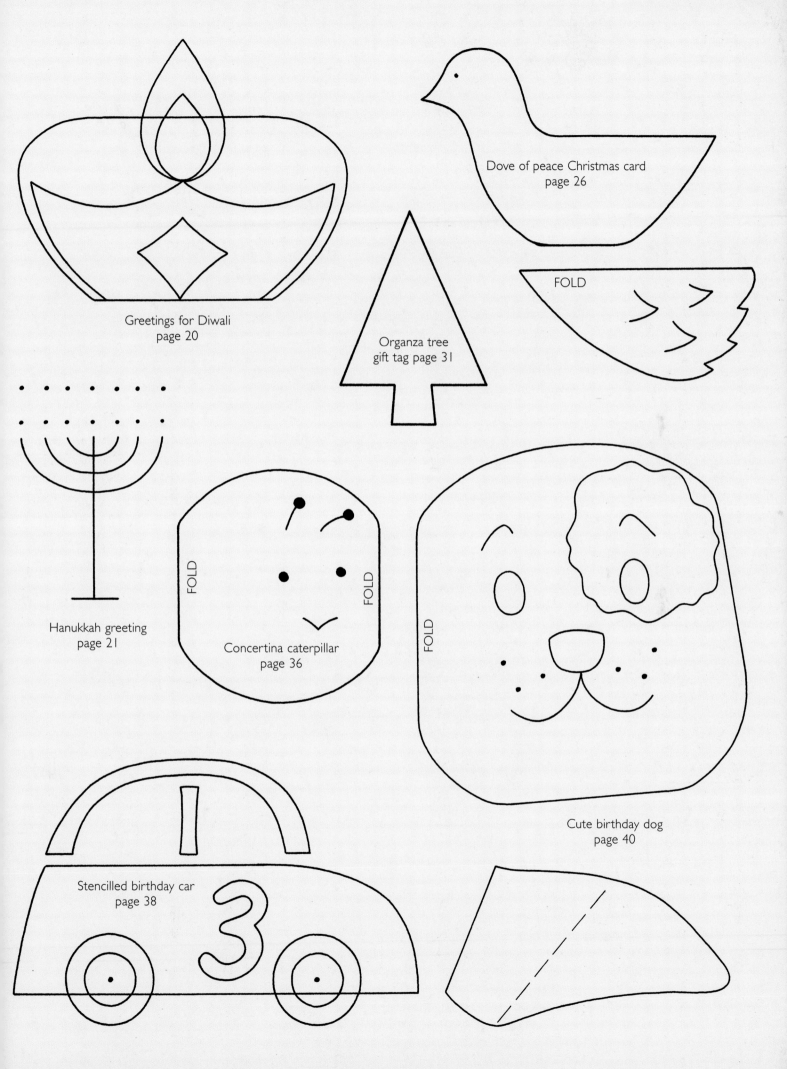

Greetings for Diwali
page 20

Dove of peace Christmas card
page 26

FOLD

Organza tree
gift tag page 31

Hanukkah greeting
page 21

FOLD   Concertina caterpillar
page 36   FOLD

FOLD   Cute birthday dog
page 40

Stencilled birthday car
page 38

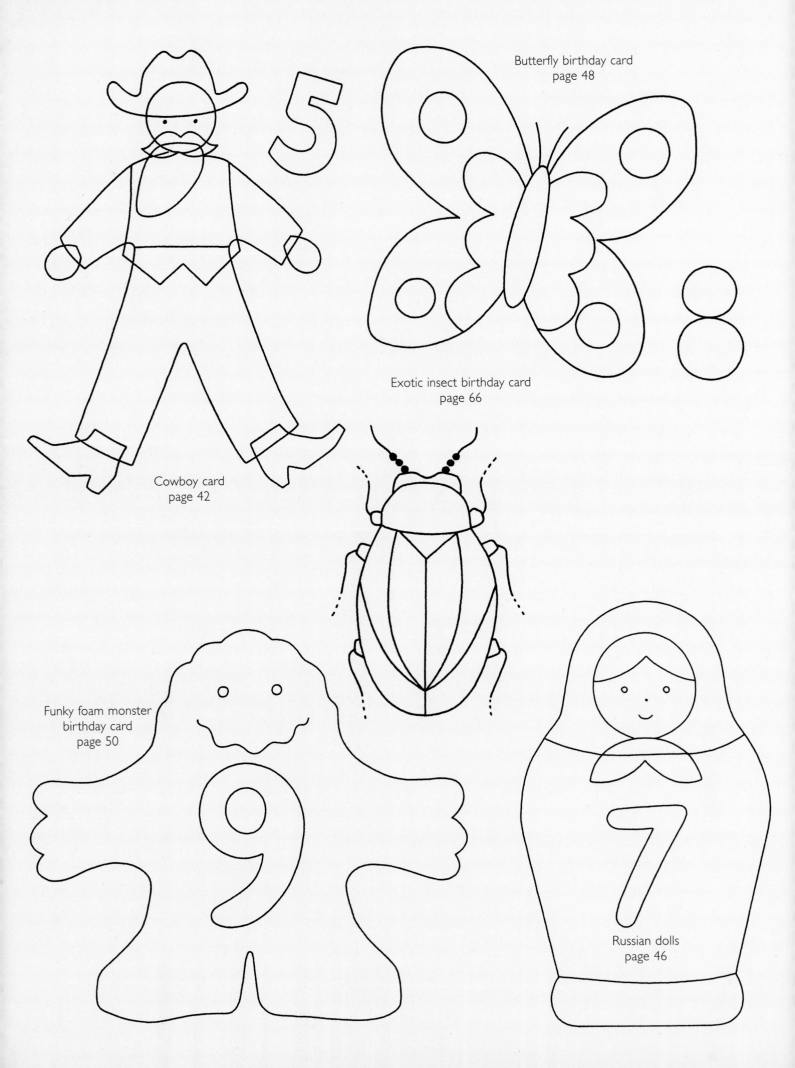

Butterfly birthday card
page 48

Exotic insect birthday card
page 66

Cowboy card
page 42

Funky foam monster
birthday card
page 50

Russian dolls
page 46

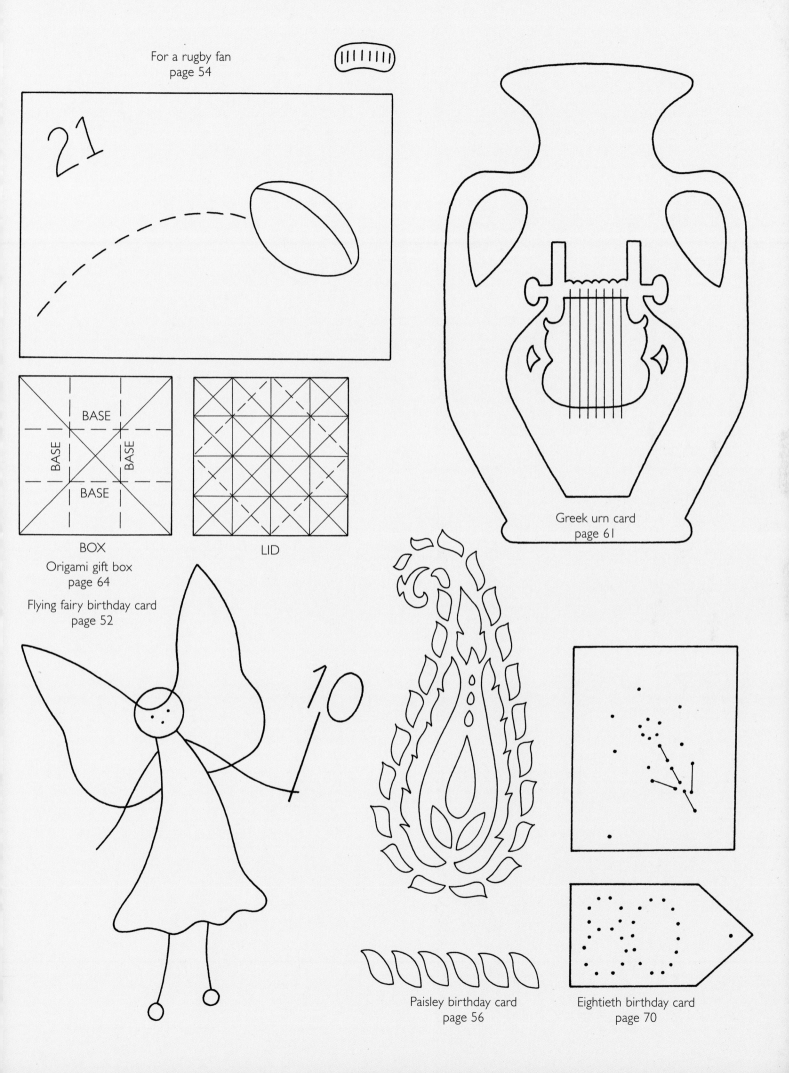

For a rugby fan
page 54

BASE
BASE
BASE
BASE

BOX

LID

Origami gift box
page 64

Flying fairy birthday card
page 52

Greek urn card
page 61

Paisley birthday card
page 56

Eightieth birthday card
page 70

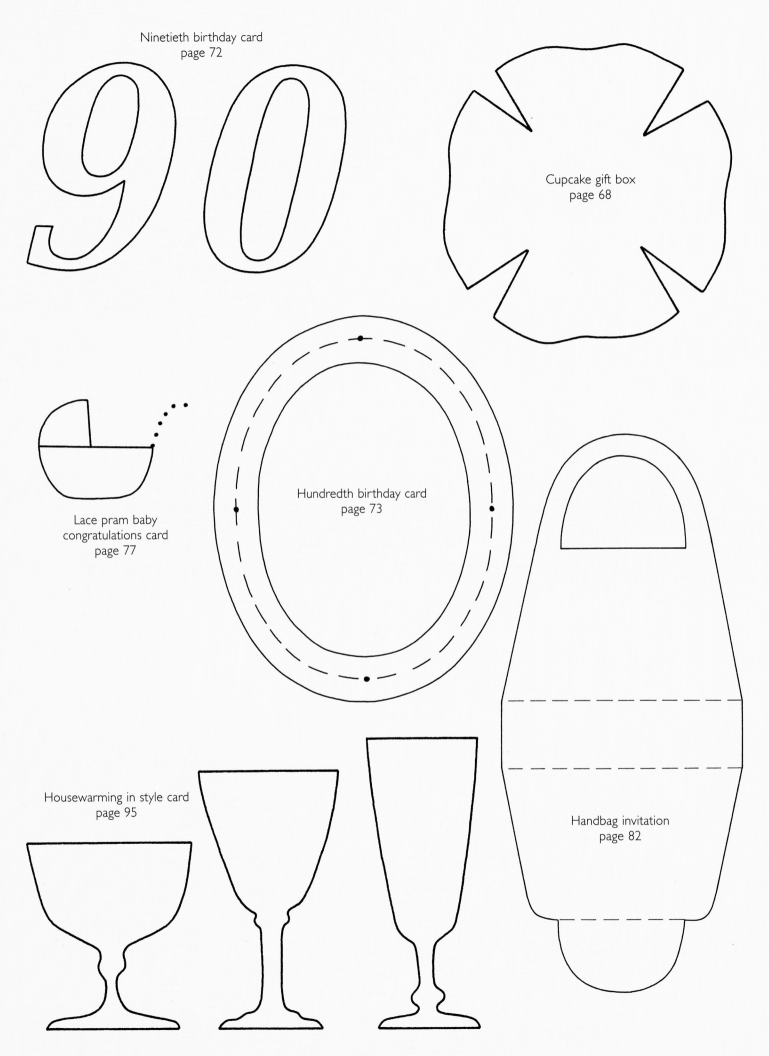

Ninetieth birthday card
page 72

Cupcake gift box
page 68

Lace pram baby
congratulations card
page 77

Hundredth birthday card
page 73

Housewarming in style card
page 95

Handbag invitation
page 82

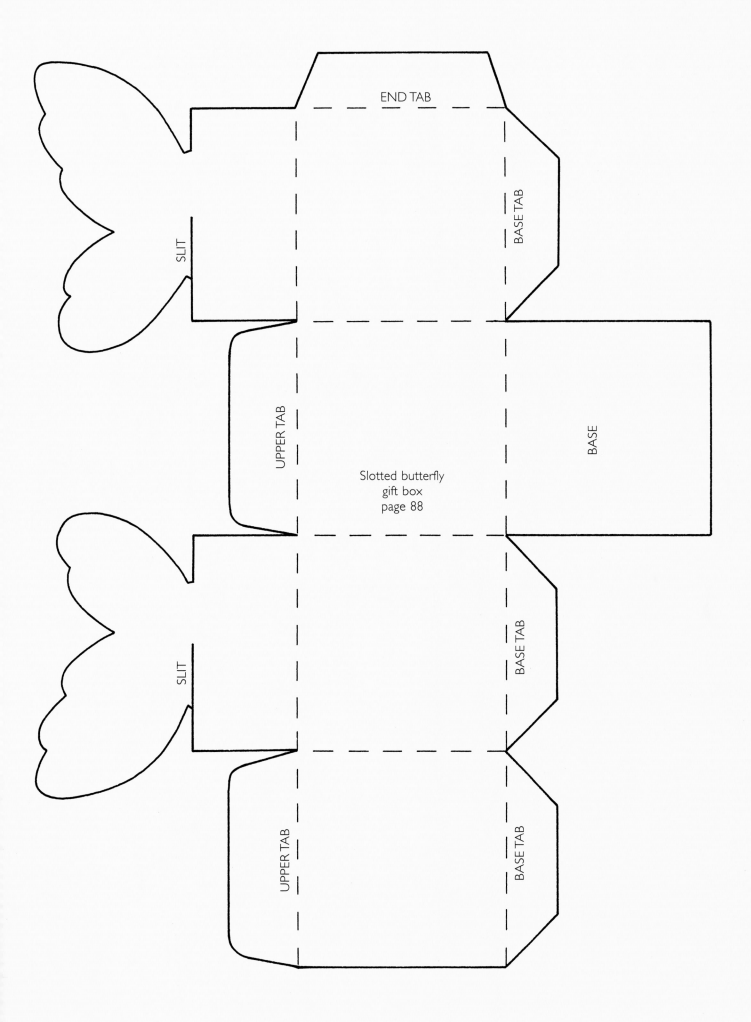

SLIT

END TAB

BASE TAB

UPPER TAB

Slotted butterfly
gift box
page 88

BASE

SLIT

BASE TAB

UPPER TAB

BASE TAB

BASE TAB

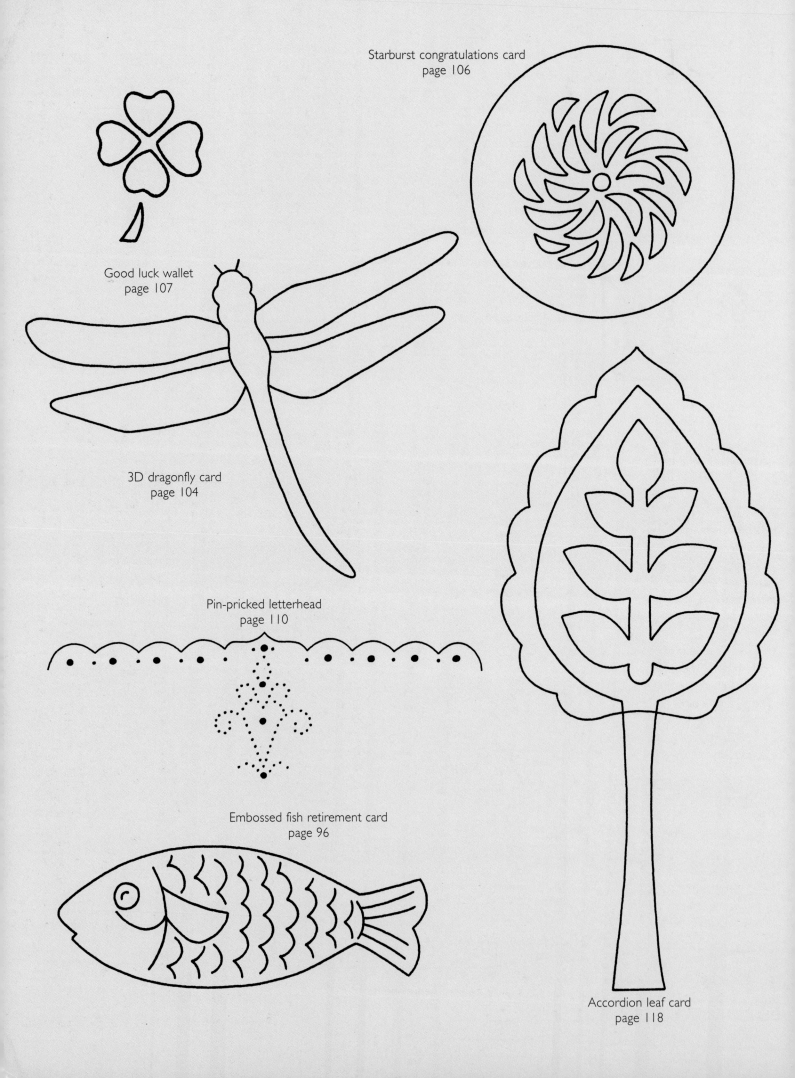

Good luck wallet
page 107

Starburst congratulations card
page 106

3D dragonfly card
page 104

Pin-pricked letterhead
page 110

Embossed fish retirement card
page 96

Accordion leaf card
page 118

INDEX